£1.50

DC CONDUCTION IN THIN FILMS

Insulating films are coming into widespread use in electronic devices and integrated circuits. As a result their academic properties are currently of great interest both in industrial research laboratories and universities. The monograph surveys the d.c. conduction mechanisms occurring in insulating films, and is written with the graduate student and industrial research worker in mind. The monograph is divided up into four chapters. The first chapter deals with the defect properties of insulators, and the metal-insulator contact, without such knowledge it is not possible to judiciously analyse experimental data. Chapter 2 deals with the various electrode limited processes, such as the tunnel effect and Schottky effects. Chapter 3 is concerned with such electrode limited processes as space charge limited currents and Poole–Frenkle effect. Chapter 4 contains a discussion of the bulk-limited to electrode-limited and electrode-limited to bulk-limited process.

Other titles in this series

EE/1 THRESHOLD LOGIC
S L Hurst

EE/2 THE MAGNETRON OSCILLATOR
E Kettlewell

EE/3 MATERIAL FOR THE GUNN EFFECT
J W Orton

EE/4 MICROCIRCUIT LEARNING COMPUTERS
I Aleksander

Other Electrical Engineering titles available

TL/EE/1 LINEAR ELECTRIC MOTORS
E R Laithewaite

TL/EE/2 AN INTRODUCTION TO THE JOSEPHSON EFFECTS
B W Petley

All published by Mills and Boon Limited

M & B Monograph EE/5

General Editor: J Gordon Cook, PhD, FRIC

DC Conduction in Thin Films

J G Simmons, BSc, PhD, FIEE, FIERE, FInstP

Professor of Electrical Engineering at the University of Toronto

Mills & Boon Limited

London

First published in Great Britain 1971
by Mills & Boon Limited, 17–19 Foley Street,
London, W1A 1DR

ISBN 0.263.51786.1

Made in Great Britain at the Pitman Press, Bath

CONTENTS

1. The Metal-Insulator-Metal System

1.1 HISTORICAL INTRODUCTION

Only in relatively recent times has it been possible to prepare thin dielectric films with any degree of reproducibility (from a conductivity standpoint). The usual means of preparation are by sputtering or vacuum deposition techniques, and even now there is a good deal of art required to obtain reproducibility. This is because the properties of the films which determine conductivity (e.g. structure, doping, etc.) are dependent on a number of factors during preparation, including residual ambient pressure, rate of deposition, substrate temperature, evaporant crucible and purity of evaporant. In other words, the conductivity of the film is extrinsic in nature. Only during recent years have vacuum coating plants and ancilliary equipment reached the level of sophistication required to permit some reasonable degree of control over the films.

In the past, experimental results have often been interpreted without due regard to the extrinsic nature of the films. A further complicating factor in interpreting experimental data is that, because of the wide range of electric fields under which the data is usually gathered, the observations often cannot be described by a single conduction process; it is not unusual for various field strength ranges to manifest different electrical phenomena. Because of the defect nature of the films, it is also not uncommon for co-operative phenomena to occur; for example, space charge (due to trapped carriers) modulation of, say, the Poole–Frenkel effect (see Section 4.1) or Schottky effect. In view of these problems, it is not surprising that the interpretation of experimental data is often contravertible. This is particularly true of the earlier data; in general, therefore, we shall be concerned primarily with discussing more recent experimental work.

1.2 TRAPPING AND IMPURITY PROBLEMS

There are several reasons for believing that the observed conductivity in vacuum-deposited thin dielectric films is often due to extrinsic properties rather than intrinsic properties. For example, the current density is often much higher than would be expected, and the activation energy associated with the conductivity is usually a good deal less than half the insulator energy gap.

The source of the extrinsic conductivity is thought to be the inherent defect nature of evaporated chemical films.[1,2] Stoichiometric films of compound insulators, with which we are primarily concerned, are notoriously difficult to prepare by evaporation, because of decomposition and preferential evaporation of the lower-vapour-pressure constituent atom. For example, using cadmium sulphide as starting material, elemental Cd tends to evaporate more rapidly and, as a result, cadmium sulphide films contain donor centres of free cadmium.[3] Silicon oxide yields a film containing a mixture of compounds varying from silicon oxide to silicon dioxide as well as free silicon.[3–6] The free silicon *per se* may act as donor centres in these films;[1] alternatively, vacancies existing in the insulator may be the source.

A good example of the existence of donors arising due to auto-doping during deposition occurs in evaporated molybdenum oxide films. The existence of the donors is readily detected by examination of capacity-frequency-temperature measurements obtained from these films.[7–9] A further problem that arises is the contamination of the films by deposits arising from sublimation of the crucible and by residual gases. Thus it requires dissociation of only one molecule per million, or the crucible to sublime at one-millionth the rate of the evaporant, to yield an impurity level of the order 10^{17} cm^{-3} within the film.

Traps are another important factor to be considered in thin films. Insulating films deposited onto amorphous (e.g. glass) substrates are usually at best polycrystalline, and in many cases amorphous. For crystallite sizes of

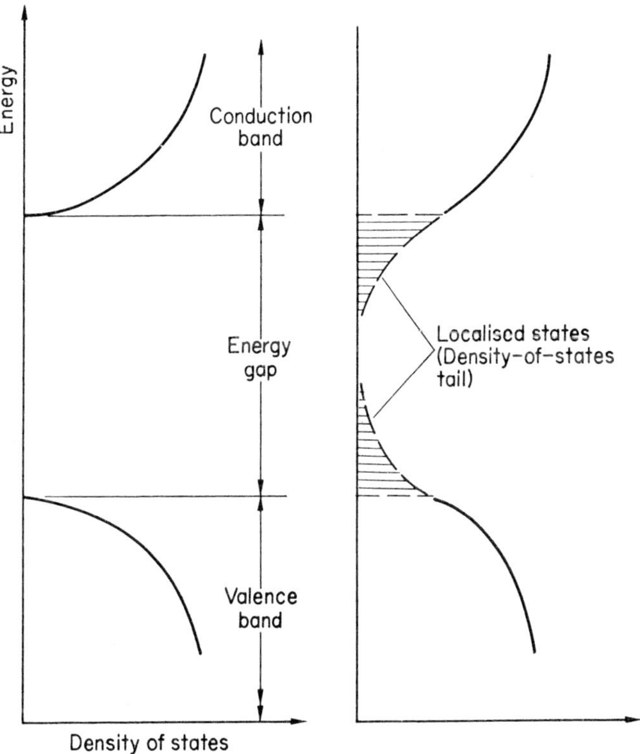

Fig. 1 Energy-density of states diagram (a) *crystalline insulator and* (b) *amorphous insulator. Note that in the case of* (a) *the conduction and valence bands and energy gap are well-defined. In* (b) *there are no such well-defined regions, and the traps arising from the "density-of-states" tail are an intrinsic property of an amorphous solid.*

100 Å, trapping densities as high as 10^{18} cm^{-3} are possible because of grain boundary defects alone; in cadmium sulphide, trapping densities as high as 10^{21} cm^{-3} have been reported.[10] Furthermore, vacuum-deposited films contain large stresses which induce further trapping centres. These traps are *extrinsic* to the insulator.

In a strictly amorphous solid there is another important consideration. Because of the lack of long-range order in these films there is a smearing of the conduction and valence band edges. Because the band edges are diffused, there is a gradual transition from a quasi-continuous state (conduction and valence bands), in which the carriers can move freely, to strictly localised states (density-of-states tail) or traps in which the carriers are immobilised[11] (see Fig. 1). These traps are an intrinsic property of amorphous insulators and probably exist in large quantities ($12 \simeq 10^{19}$ cm^{-3}).

It follows that thin film vacuum-deposited insulators can contain a large density of both impurity and trapping centres. A judicious study of electrical conduction in vacuum-deposited thin films cannot be accomplished without consideration of these possibilities.

1.3 THE INSULATOR IN ISOLATION (ELECTRODELESS INSULATOR)

In a perfect (trap-free) insulator, the position of the intrinsic Fermi level E_{Fi} is given by (see Fig. 2(a)):

$$E_{Fi} = \frac{E_g}{2} + \frac{kT}{2} \ln \left(\frac{N_v}{N_c}\right) \qquad (1)$$

where N_c and N_v are respectively the effective density of states in the conduction and valence bands, E_c and E_v are respectively the energies of the bottom of the conduction and valence bands, E_g is the insulator energy gap, k is

Boltzmann's constant and T is the absolute temperature. Normally, $(kT/2)\ \ln\ (N_v/N_c)$ is small compared to $E_g/2$ so that the intrinsic Fermi level lies close to the middle of the energy gap.

In practice, an insulator abounds with traps. Traps positioned above the Fermi level are essentially empty (shallow) while those positioned below the Fermi level are essentially filled (deep).(13) Consider an insulator containing a single discrete shallow and a single discrete deep trap level* (Fig. 2(b)). Non-degenerate equilibrium statistics require that

$$N_{td} \exp\left(\frac{E_{td} - E_{Fi}}{kT}\right) + N_v \exp\left(\frac{E_v - E_{Fi}}{kT}\right)$$
$$= N_{ts} \exp\left(\frac{E_{Fi} - E_{ts}}{kT}\right) + N_c \exp\left(\frac{E_{Fi} - E_c}{kT}\right) \qquad (2)$$

where N_{td} and N_{ts} are respectively the deep and shallow trap densities and E_{td} and E_{ts} are respectively the energies of the deep and shallow trap levels. For practical values of trap parameters we may neglect the two terms involving the conduction and valence band, and from the remaining two terms obtain the insulator Fermi energy:

$$E_{Fi} = \frac{E_{ts} + E_{td}}{2} + \frac{kT}{2} \ln\left(\frac{N_{td}}{N_{ts}}\right) \qquad (3)$$

From equation (3) it is seen that E_{Fi} is determined only by the trap parameters, and it is normally positioned between the two trap levels, as shown in Fig. 2(b).

* Because of limited space, discussion is confined to this type of system. Actually, the distributed-traps case is probably the more important from a practical sense, but it is too complicated a problem to treat at this level. (See J. G. Simmons, *Physics and Chemistry of Solids* (to be published).) The treatment given here is provided to illustrate the importance of trapping levels.

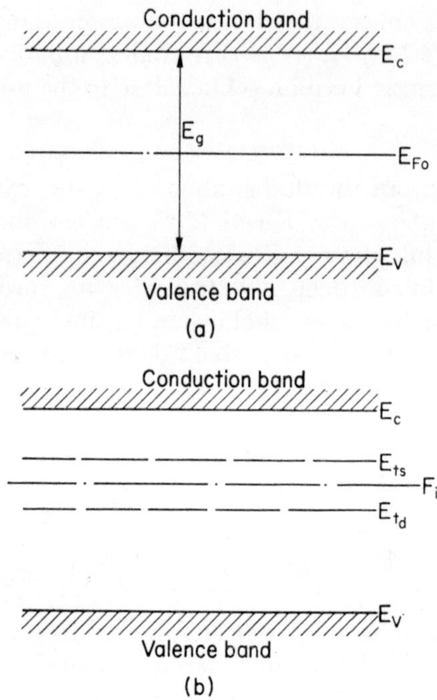

Fig. 2 Energy diagram of an insulator (a) *which is trap-free and* (b) *having a discrete shallow trap and a discrete deep trap.*

1.4 METAL-INSULATOR-METAL SYSTEM

When electrodes are applied to the insulator, the underlying principle in determining the height of the potential barrier and the shape of the bottom of the insulator conduction band is that the vacuum and Fermi levels of the electrode and insulator are continuous across the interface. Actually, now we do not distinguish the electrode and insulator Fermi level; they are one and the same level: the *system* Fermi level, E_{Fs}. The continuity of the vacuum levels determines the height of the interfacial barrier ϕ_0 (see Fig. 3):

$$\phi_0 = \psi_m - \chi$$

where ψ_m is the metal work function and χ the insulator affinity.

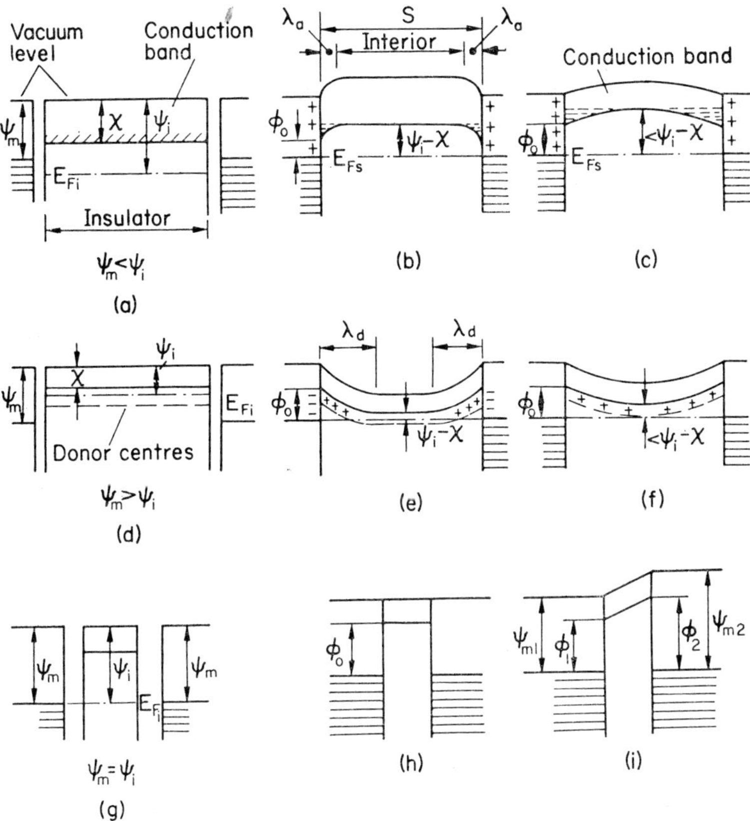

Fig. 3 Energy diagrams of metal-insulator-metal systems having (a)–(c) *ohmic contact;* (d)–(f) *blocking contact;* (g)–(h) *neutral contact,* (i) *see text.*

Normally, before the electrodes are applied, $\psi_m \neq \psi_i$ (where ψ_i is the *original* work function of the insulator). Thus, when the electrodes are initially applied, charge is exchanged between the electrode and insulator such that the change in position of the insulator Fermi level (resulting from the change in occupancy of the electron traps in the insulator) reduces the energy differential between the electrode and insulator Fermi levels. The charge exchange ceases when the two Fermi levels coalesce to form the discrete monoenergetic system Fermi level as shown in Fig. 3.

The types of contact that can exist at a metal-insulator interface fall into three categories:[14] (1) ohmic contact, (2) neutral contact, and (3) blocking contact. Each of the contacts will now be considered. Our remarks will be confined to the case of shallow trapping; the reader is referred to the literature for the case of deep traps.[13]

(a) OHMIC CONTACT

In order to achieve an ohmic contact at a metal-insulator interface it is necessary that the electrode work function, ψ_m, be smaller than the insulator work function, ψ_i, as shown in Fig. 3(a). Under this condition, in order to satisfy the thermal equilibrium requirements, electrons are injected from the electrode into the conduction band of the insulator, thus giving rise to a space charge region, and hence a space-charge-induced field in the insulator. This field causes the bottom of the conduction band to curve upwards away from the interface. The width of the space charge region λ_a is the distance from the electrode-insulator interface required for the bottom of the conduction band to rise an energy $\psi_i - \chi$ above the system Fermi level (or an energy $\psi_i - \psi_m$ above the top of the interfacial barrier (see Fig. 3(b)). This space-charge region is termed the *accumulation* region. The width of the accumulation region, λ_a, when only a discrete shallow trapping level is present in the insulator, is given by[14]

$$\lambda_a = \frac{\pi}{2}\left(\frac{2kT}{e^2 N_t} K\varepsilon_0\right)^{\frac{1}{2}} \exp\left(\frac{E_{ts} - E_{Fs}}{kT}\right)^{*} \tag{4}$$

(For the expressions for λ_a when deep trapping occurs the reader is referred to the literature.)[14] Figure 4 shows a plot of λ_a versus $E_{ts} - E_{Fs}$ for various N_t, using K = 5.

* Where ε_0 is the permittivity of free space and K is dielectric constant of the insulator.

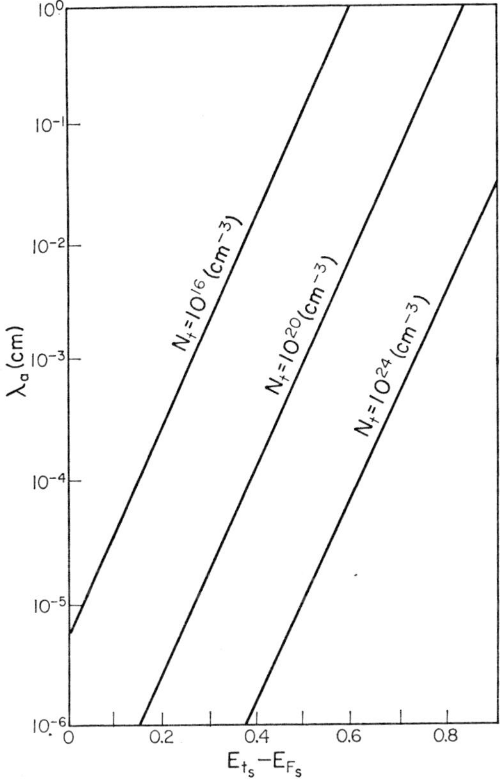

Fig. 4 Size of the accumulation region, λ_a, as a function of $E_{ts}-E_{Fs}$.

Figure 3(a–c) illustrates the energy diagram for two ohmic contacts at the same potential. Figure 3(b) illustrates the case of non-overlapping accumulation regions ($S > 2\lambda$, where S is the insulator thickness). In this case the accumulation regions effectively screen the interior from conditions at the surface, and the bottom of the conduction band between the accumulation regions is flat (space charge and field free). The only charge contained in the interior is the indigenous charge, which means that the height of the conduction band is $\psi_i - \chi$ above the Fermi level.

On the other hand, Fig. 3(c) illustrates the case of overlapping accumulation regions (that is $S < 2\lambda$), in which case space charge extends throughout the insulator. This occurs when there is insufficient charge within the surface of the insulator to screen the interior effectively from conditions at the interfaces. A consequence of the overlapping accumulation regions is that the highest point of the bottom of the insulator conduction band is less than $\psi_i - \chi$ above the Fermi level.

The ohmic contact, which acts as a reservoir of charge, is capable of supplying electrons to the insulator, as required by bias conditions. Thus, with contacts of this type, the conduction process through the system is limited by the rate at which the electrons can flow through the bulk of the insulator, rather than the rate at which they are supplied by the electrode; hence, the conduction process is *bulk* limited.

(b) BLOCKING CONTACT–SCHOTTKY BARRIER

A blocking contact occurs when $\psi_m > \psi_i$ (see Fig. 3(d)), and in this case electrons flow from the insulator into the metal to establish thermal-equilibrium conditions. A space-charge region of positive charge, the depletion region, is thus created in the insulator, and an equal negative charge resides on the metal electrode. As a result of the electrostatic interaction between the oppositely charged regions, a local field exists within the surface of the insulator. This causes the bottom of the insulator conduction band within the bulk of the insulator to bend downwards until it lies an energy $(\psi_i - \chi)$ above the system Fermi level (or $\psi_m - \psi_i$ below the top of the interfacial barrier); the plane at which this occurs determines the edge of the depletion region. Assuming a uniform donor density of N_d within the insulator* (see Section 1.2)

* The insulator is required to contain a relatively high density of deep traps, the energy of which must satisfy $(E_c - E_{td} + \chi) < \psi_m$, or donor centres if λ_d is to be less than 1μ.

we have for the width of the depletion region λ_d

$$\lambda_d = \left[\frac{2(\psi_m - \psi_i)\, K\varepsilon_o}{e^2 N_d}\right]^{1/2} \tag{5}$$

Table 1 gives a few values for λ_o using $\psi_m - \psi_i = 2$ eV and $K = 5$.

Table 1. Depth of Depletion Region for Several Values of N_d

N_d cm^{-3}	10^{15}	10^{17}	10^{19}	10^{21}
λ_d cm	10^{-4}	10^{-5}	10^{-6}	10^{-7}

Figures 3(e) and (f) illustrate two blocking contacts on a doped insulator. In Fig. 3(e) the depletion regions effectively screen the interior from the surfaces ($S > 2\lambda_d$). Under these conditions the interior of the insulator is field-free, and the bottom of the conduction band is positioned $\psi_i - \chi$ above the system Fermi level within the interior. On the other hand, in Fig. 3(f) the depletion regions overlap ($L > 2\lambda$). As in the case of Fig. 3(c), the conduction band is bent throughout its length, but it is concave upwards, reflecting that the space charge is positive, in contrast to that of Fig. 3(c) in which it is negative. Hence the depletion regions do not contain sufficient charge to screen the interior from the surface; thus the bottom of the conduction band is *greater* than $\psi_i - \chi$ above the Fermi level at the centre of the insulator.

It will be apparent from the energy diagram that the free electron density at the interface is much lower than that in the bulk of the insulator. Thus the rate of flow of electrons through the system will be limited by the rate at which they flow over the interfacial barrier; hence the conduction process is electrode limited.

(c) NEUTRAL CONTACT

When $\psi_m = \psi_i$ the vacuum and Fermi levels of the insulator and electrode line up naturally without the necessity of charge transfer between the electrode and insulator. Since no space charge exists within the insulator, the conduction band is flat right up to the interface; that is, no band bending is present, as shown in Fig. 3(h). This contact is known as a neutral contact and represents the transitional stage between an ohmic and blocking contact. This type of contact also pervails if $\psi_m < \psi_i$ and the trap level is positioned more than about 1 eV above the system Fermi level. In this case, for practical trap densities, the amount of space charge contained in the traps is too small to give rise to significant band bending.(14) For initial voltage bias, the cathode is capable of supplying sufficient current to balance that flowing in the insulator, and, since the field in the insulator is constant, the conduction process is ohmic. There is, however, a limit to the current that the cathode can supply, and this is the saturated thermionic (Richardson) current over the barrier. When this limit is reached, the conduction process ceases to be ohmic and becomes electrode limited (see Section 2.2). When the insulator has dissimilar electrodes connected to its surfaces, it is clear that the interfacial potential barriers differ in energy by an amount

$$(\psi_{m2} - \chi) - (\psi_{m1} - \chi) = (\psi_{m2} - \psi_{m1})$$

as shown in Fig. 3(i). If the traps are positioned sufficiently high above the system Fermi level so that there is negligible charge trapped in the insulator to cause significant band bending, a uniform intrinsic field F_{in} of strength $(\psi_{m2} - \psi_{m1})/es$ exists within the insulator. The origin of this zero-bias intrinsic field is a consequence of charge transfer *between* the electrodes. The electrode of lower work function, electrode 1, transfers electrons to electrode 2, so that a positive surface charge appears on electrode 1 and a negative surface charge on electrode 2. The amount of charge Q transferred between the electrodes (the surface charge on the electrodes) is:

$$Q = \frac{(\psi_{m2} - \psi_{m1}) A K \varepsilon_0}{es}$$

where A is the electrode area.

If the insulator is very thin, the intrinsic field within the insulator can be very large; for example, suppose $s = 20$ Å, as one finds in metal-insulator-metal tunnel junctions (see Section 2.1), and $(\psi_{m2} - \psi_{m1})/e = 1$ V, then $F_{in} = 5 \times 10^6$ V cm^{-1}. By investigating the breakdown voltage of very thin metal-insulator-metal junctions as a function of voltage bias, the presence of the intrinsic field can be detected.(15)

1.5 EFFECT OF SURFACE STATES

In the above discussion, we have assumed the barrier height at the metal-insulator interface to be $\psi_m - \chi$. If surface states exist on the insulator surface, this is not necessarily true;(16) in fact, if the surface density is high ($\simeq 10^{13}$ cm^{-3} eV^{-1}), the interfacial barrier height is virtually independent of the electrode material. These states owe their existence in part to the sudden departure from periodicity of the potential at the surface (Tamm States) and in part to the defect chemical nature of the surface, i.e. absorbed gas, etc., or in other words, to the manner in which the film is prepared. Thus, it is normally not possible to determine *a priori* if surface states will exhibit a dominant role in the determination of the interfacial barrier height. This being the case, throughout the rest of this article we will assign the parameter ϕ to the interfacial height, and from the foregoing its value will depend on whether or not surface states exist at the metal-insulator interface.

2. Electrode Limited Processes

The electron current flowing between two metal electrodes separated by an insulator is given by[17,18]

$$J = (4\pi\ em/h^3) \int_0^\infty dE\ [f_c(E) - f_a(E)] \int_0^E P(E_x)\ dE_x \quad (6)$$

where e = unit of electronic charge; h = Planck's constant;

$f_c = [1 + e^{(E-E_f)/kT}]^{-1}$ and $f_a = [1 + e^{(E+eV-E_f)/kT}]^{-1}$

are respectively the electron distributions in the cathode and anode electrodes; E_f is the Fermi energy of the cathode; V is the voltage difference between the cathode and anode; E and E_x are respectively the energy and "x-directed" energy of the electron; and $P(E_x)$ is the electron transmission probability through the insulator. Equation (6) assumes the parabolic electron energy-momentum relation and the free electron mass, m, in each of the three regions of space. If the insulator is sufficiently thin, the predominant contribution to the current is derived from electrons in the cathode with $E_x \simeq E_f$ tunnelling through the insulator potential barrier. For thicker films and higher temperatures the current will be due to electrons thermally excited over the insulator potential barrier (thermionic emission—Section 2.2). Each of these processes in turn.

2.1 TUNNEL EFFECT

(a) THERMAL J–V CHARACTERISTIC

Stratton[18] first derived the thermal J–V characteristic for a tunnel junction using equation (6) and the WKB approximation for $P(E_x)$. However, the simplest generalised formulation connecting the tunnel current $J(V,T)$ at T°K to the tunnel current $J(V,0)$ at 0°K is given by[19]

$$\frac{J(V,T)}{J(V,0)} = \frac{(\pi\ AkT/2\bar{\phi}^{1/2})}{\sin\ (\pi\ AkT/2\bar{\phi}^{1/2})} \tag{7}$$

where

$$J(V,0) = [e/2\pi\ h(\Delta s)^2]\{\bar{\phi}\ \exp\ (-A\bar{\phi}^{1/2}) - (\bar{\phi} + eV)\ \exp\ [-A(\bar{\phi} + eV)^{1/2}]\} \tag{8}$$

and $A = 4\pi\ \Delta s(2m)^{1/2}/h$, Δs = width of the potential barrier at the Fermi level of the cathode and $\bar{\phi}$ = mean height barrier height above the Fermi level of the cathode (see Fig. 5). With $\bar{\phi}$ expressed in electron volts and Δs in angstrom units

$$\frac{J(V,T)}{J(V,0)} \simeq 1 + \frac{3 \times 10^{-9}(\Delta s T)^2}{\bar{\phi}} \tag{9}$$

Using typical values of $\bar{\phi}$ (=2 eV) and Δs(=20 Å) we have

$$J(V,T) = J(V,0)\ (1 + 6 \times 10^{-7}T^2) \tag{10}$$

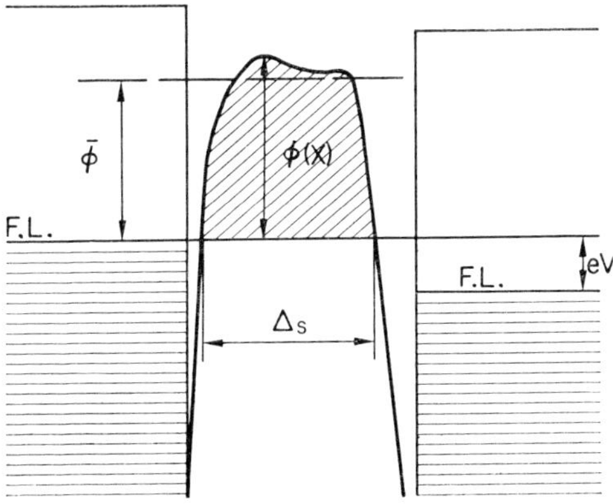

Fig. 5 Energy diagram of an arbitrary barrier illustrating the parameters $\bar{\phi}$ *and* Δs.

This slight quadratic dependence of the current (at constant voltage) on temperature is characteristic of the tunnel process.

An interesting result is apparent when we rewrite equation (9) in the following form:

$$\begin{aligned} \hat{J} &= 100\{[J(V,T)-J(V,0)]/J(V,0)\} \\ &= 3 \times 10^{-7}\,(\Delta s T)^2/\bar{\phi} \end{aligned} \tag{11}$$

$\hat{J}$ will be recognised as the percentage change in current (for a fixed voltage bias) as the temperature is raised from 0°K to T°K. For a trapezoid barrier of the type shown in Fig. 3(i), we have from Figs. 6(a) and 7(a), for reverse bias:

$$\hat{J}_1 = \begin{cases} 6 \times 10^{-7}\,(sT)^2/(\phi_1 + \phi_2 - V);\ V \leq \phi_2 \\ 6 \times 10^{-7}\,\phi_1\,[sT/(V - \Delta\phi)]^2;\ V \geq \phi_2 \end{cases} \tag{12}$$

and from Figs. 6(b) and 7(b), for forward bias:

$$\hat{J}_2 = \begin{cases} 6 \times 10^{-7}\,(sT)^2/(\phi_1 + \phi_2 - V);\ V < \phi_1 \\ 6 \times 10^{-7}\,\phi_2[sT/(V + \Delta\phi)]^2;\ V > \phi_1 \end{cases} \tag{13}$$

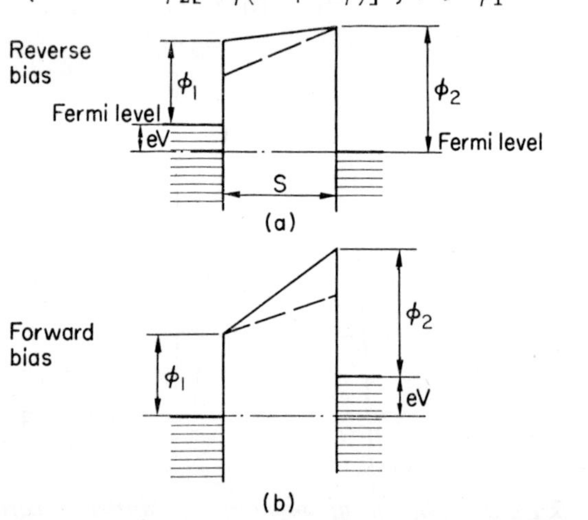

Fig. 6 Energy diagrams of asymmetric potential barrier for $0 < V < \phi_1/e$*:* (a) *is reverse-bias condition, and* (b) *is forward-bias condition. The dotted lines illustrate the shape of the barrier (with respect to the Fermi level of the positively-biased electrode in both cases) under zero bias condition (cf. Fig. 3(i)).*

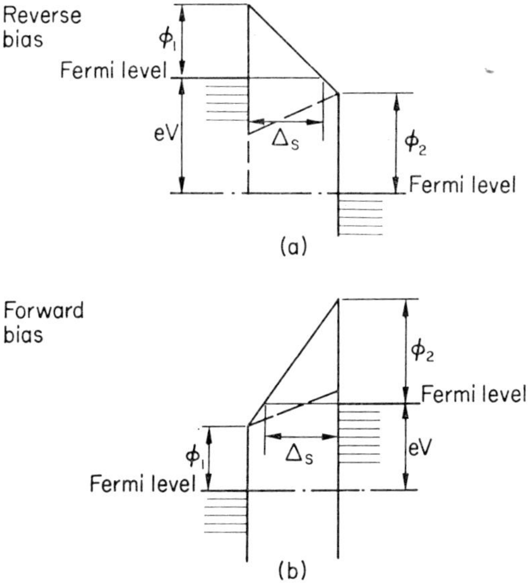

Fig. 7 Energy diagrams of potential barriers for $V > \phi_2/e$ (see Fig. 6 for explanation of dotted lines).

Plots of $\hat{J}_1$ and $\hat{J}_2$ versus V are shown in Fig. 8. These curves peak sharply at $V = \phi_1$ and $V = \phi_2$ for forward and reverse biases respectively, and thus provide a sensitive method of determining ϕ_1 and ϕ_2.

The very thin insulating films required to observe the thermal tunnel effects described above are usually the thermally[20] or anodically grown[21] surface oxides of one of the electrodes. These films can be grown pin-hole free to thicknesses as low as 15 Å. Thus the usual method of preparing a tunnel junction is to grow the insulating film on the surface of a vacuum-deposited thin film of, say, aluminium (parent electrode). The junction is then completed by depositing a counterelectrode on to the free surface of the oxide layer.

The $\hat{J}$–V data of Pollack and Morris[22] for Al–Al_2O_3–Al junctions in which the aluminium oxide was anodically grown are transcribed in Fig. 9. The similarity between these curves and the theoretical $\hat{J}$–V curves in Fig. 8 is unmistakable, but the percentage increase in the current is greater than theoretically expected. The barrier heights determined from this data, that is, the voltage at which the current peaks occur, are $\phi_1 = 2{\cdot}0$ eV and $\phi_2 = 2{\cdot}2$ eV.

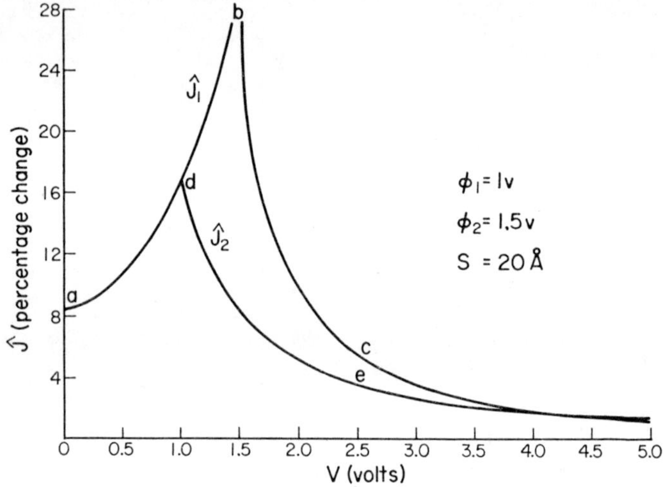

Fig. 8 Graph of $\hat{J}$ versus V ($T = 300°K$). $\hat{J}$ represents the percentage change in current, at a voltage bias V, as the temperature of the tunnel junction is raised from 0°K to 300°K. Curve abe *is the forward-biased characteristic ($\hat{J}_2$–V) and* adc *the reverse-biased characteristic ($\hat{J}_1 V$).*

It is not too surprising that the two barrier heights are not identical in what is an *apparently* symmetrical system. This is because the interface between the parent metal and its grown oxide probably contains a semiconducting transition region connecting the metallic and oxide regions. On the other hand, the interface between the counterelectrode and the free surface of the oxide probably consists of a relatively sharp transition between the metallic and insulating regions. The $\hat{J}$–V characteristic for an Al–Al_2O_3–Au junction is also shown in Fig. 9, but the peak in this characteristic is not as sharp as for the Al–Al_2O_3–Al junction. The displacement between the peak of this characteristic and that of the corresponding Al electrode of the Al–Al_2O_3–Al junction is 0·6 V, which is in reasonable agreement with the difference in work functions (0·75 eV) of aluminium and gold.

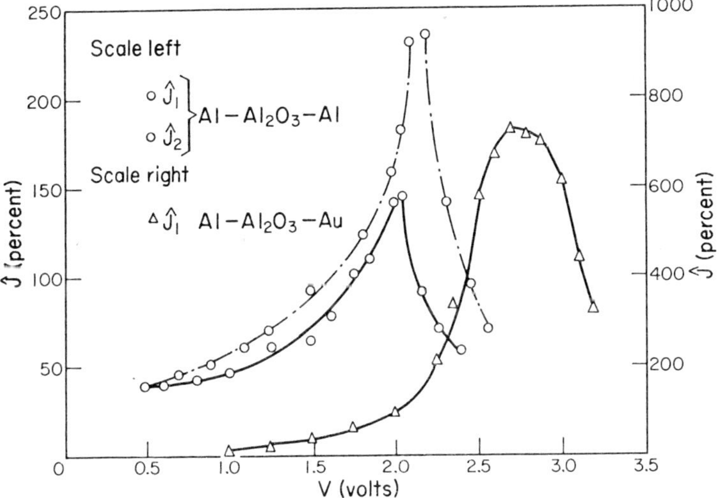

Fig. 9 $\hat{J}$–V tunnel characteristic for Al–Al_2O_3–Al and Al–Al_2O_3–Au junction.

(b) ISOTHERMAL J–V CHARACTERISTIC

Since the tunnel currents are only very slightly temperature dependent, the expression $J(V,0)$ may be used to describe the *isothermal* J–V characteristic at any reasonable temperature. This is a particularly expeditious approximation to use when analysing experimental tunnel data, because of the simpler functional form of $J(V,0)$ compared to $J(V,T)$. Expressing Δs and $\bar{\phi}$ in angstroms and electron volts, respectively, and J in A cm^{-2} equation (8) reduces to[23,24]

$$J = \frac{6{\cdot}2 \times 10^{10}}{(\Delta s)^2} \{\bar{\phi} \exp(-1{\cdot}025\, \Delta s \bar{\phi}^{1/2}) - (\bar{\phi} + V) \exp[-1{\cdot}025\, s(\bar{\phi} + V)^{1/2}]\} \quad (14)$$

Using $\bar{\phi}$ and Δs from Fig. 6 in equation (14) we have (for a trapezoidal barrier) for *both forward* (J_2) and *reverse* (J_1) currents

$$J_1 = J_2 = (3{\cdot}1 \times 10^{10}/s^2)\{(\phi_1 + \phi_2 - V) \times \exp[-0{\cdot}725 s(\phi_1 + \phi_2 - V)^{1/2}] - (\phi_1 + \phi_2 + V) \times \exp[-0{\cdot}725 s(\phi_1 + \phi_2 + V)^{1/2}]\};\ 0 < V \leq \phi_1 \quad (15)$$

Since $J_1 = J_2$ the J–V characteristic is symmetric (see, however, Rowell[25]) with polarity of bias for $0 < V \leq \phi_1$.

Using $\bar{\phi}$ and Δs from Fig. 7(a) in equation (16) we have for the reverse-biased condition, using $\Delta\phi = (\phi_2 - \phi_1)$:

$$J_1 = \frac{3{\cdot}38 \times 10^{10}(V - \Delta\phi)^2}{\phi_1 s^2}\left[\exp\left(-\frac{0{\cdot}69 s \phi_1{}^{3/2}}{V - \Delta\phi}\right) - \left(1 + \frac{2V}{\phi_1}\right)\exp\left(-\frac{0{\cdot}69 s \phi_1{}^{3/2}(1 + 2V/\phi_1)^{1/2}}{V - \Delta\phi}\right)\right];\ V > \phi_2 \quad (16)$$

and from Fig. 7(b) for the forward-biased condition

$$J_2 = \frac{3{\cdot}38 \times 10^{10}(V + \Delta\phi)^2}{\phi_2 s^2}\left[\exp\left(-\frac{0{\cdot}69 s \phi_2{}^{3/2}}{V + \Delta\phi}\right) - \left(1 + \frac{2V}{\phi_2}\right)\exp\left(-\frac{0{\cdot}69 s \phi_2{}^{3/2}(1 + 2V/\phi_2)^{1/2}}{V + \Delta\phi}\right)\right];\ V > \phi_1 \quad (17)$$

In this case, equations (16) and (17) are not equivalent. It follows, then, that the J–V characteristic is asymmetric in this range. In actual fact, not only is the J–V characteristic asymmetric with polarity of bias, but also the direction of easy conductance reverses at some particular voltage, as shown in Fig. 10. In Fig. 10 the tunnel resistance ($=V/J$) versus V is plotted rather than J versus V, because it is a more efficacious way of illustrating the effect of the junction parameters on the tunnel characteristics.

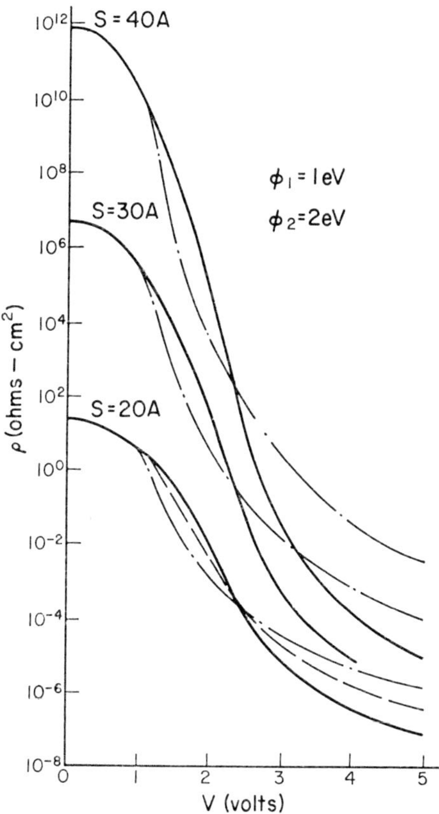

Fig. 10 $\rho(=J/V)$–V tunnel characteristics for $\phi_1 = 1eV$, $\phi_2 = 2eV$ and $s = 20$, 30 and 40 A.

Figure 11 illustrates the isothermal tunnel of Pollack and Morris[26] obtained from Al–Al_2O_3–Al tunnel junctions in which the oxide was thermally grown on the parent metal. They observed correlation between theory and experiment over nine decades of current density, and determined the interfacial barrier heights to be 1·6 and 2·5 eV. Again, the difference in barrier heights for an apparently symmetrical system reflects the difference in the two metal-insulator interfaces, as previously mentioned in connection with Fig. 9.

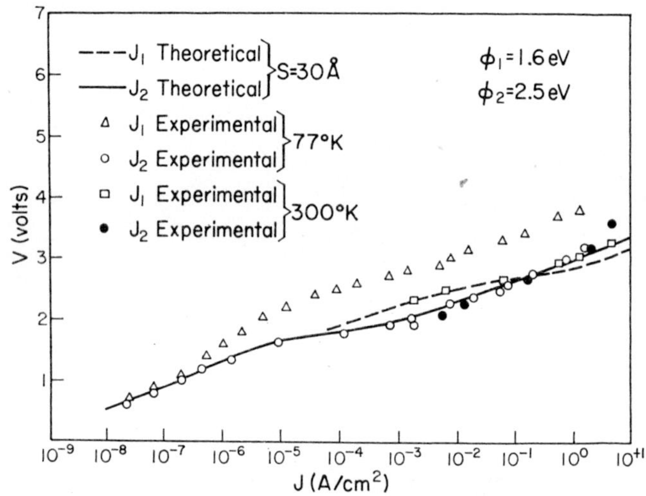

Fig. 11 Tunnel data from Al–Al_2O_3–Al junction.

It is readily shown from equations (14) and (16), and equations (15) and (17) that the logarithmic derivative of the tunnel current with respect to voltage (dJ/dV) versus V peaks sharply at $V = \phi$, and $V = \phi_2$ for forward and reverse bias respectively.[27] This technique provides a more sensitive and less laborious method of determining barrier heights than that of adjusting the barriers to obtain the best fit of the theory to the experimental isothermal J–V characteristic, as used by Pollack and

Morris. Gundlach and Heldman[28] used this technique to determine the barrier heights in tunnel junctions prepared in similar manner to those used by Pollack and Morris[26] as described immediately above. From their data they obtained barrier heights of 1·6 and 2·4 eV, which are in excellent agreement with those obtained by Pollack and Morris.

(c) FOWLER–NORDHEIM TYPE TUNNELLING

At high voltages, i.e., $V \gg \phi_2/e$, both equations (16) and (17) reduce to the familiar Fowler–Nordheim[29] form:

$$J = \frac{3{\cdot}38 \times 10^{10} F^2}{\phi} \exp\left(-\frac{0{\cdot}69\phi^{3/2}}{F}\right) \qquad (18)$$

Since the exponent is the dominant variable in equation (18), then a plot of ln J versus F^{-1}, yields a straight line of scope $(-0{\cdot}69\phi^{3/2})$.

Lenzlinger and Snow[30] studied tunnelling in the silicon-silicon oxide–aluminium/magnesium system in which the oxide was relatively thick (600–5,000 Å). Thus it was necessary to apply high voltage biases to the system to reduce the barrier thickness at the negatively-biased Fermi level to a sufficiently small value (<50 Å) in order to observe the (Fowler–Nordheim type) tunnel effect. Fowler–Nordheim characteristics were observed (see Fig. 12) over more than five decades of current emission, but the absolute values of current were lower by a factor of five to ten than the theoretically expected values, probably due to trapping effects. From the slopes of the Fowler–Nordheim characteristics (log J/F^2 versus F^{-1}; cf. equation (18)), the relative effective electron mass in the forbidden band of the silicon dioxide was found to be about 0·4*. The temperature dependence of the current was found to follow the theoretical curve from 80°–420°K, but it was necessary to assume an effective

* For an effective electron mass m^* other than unity the exponent in equation (18) becomes $-(0{\cdot}69\phi^{3/2}/F)\,(m^*/m)$.

mass of 0·95, which is inconsistent with the value obtained from the isothermal data. They suggest that the anomaly could be resolved if the interfacial barrier height *per se* was temperature dependent.

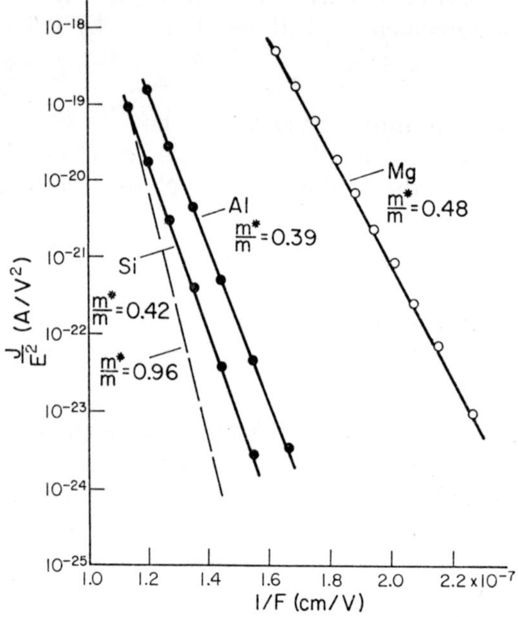

Fig. 12 Fowler–Nordheim characteristic for Si–SiO_2–Al/Mg system. The values of the relative effective mass m/m corresponding to the respective slopes are indicated. Dotted line is the slope corresponding to a $(m^*/m) = 0{\cdot}96$, as obtained from current versus temperature measurements. Negatively-biased electrode (Si, Al, Mg) marked alongside curves.*

(d) NEGATIVE RESISTANCE EFFECTS IN METAL-OXIDE-SEMICONDUCTOR TUNNEL JUNCTIONS

Esaki and Stiles[(31)] fabricated Al–Al_2O_3–SnTe junctions by evaporating stannic telluride on to an oxidised strip of aluminium. The stannic telluride layer was highly *p*-type with about 8×10^{20} carriers cm^{-3}. The *J–V* characteristics of these junctions (aluminium electrode negatively biased) taken at $T = 4{\cdot}2$, 77 and 300°K, are shown in Fig. 13. The interesting feature of these curves is the negative resistance region in the range 0·55–0·85 eV.

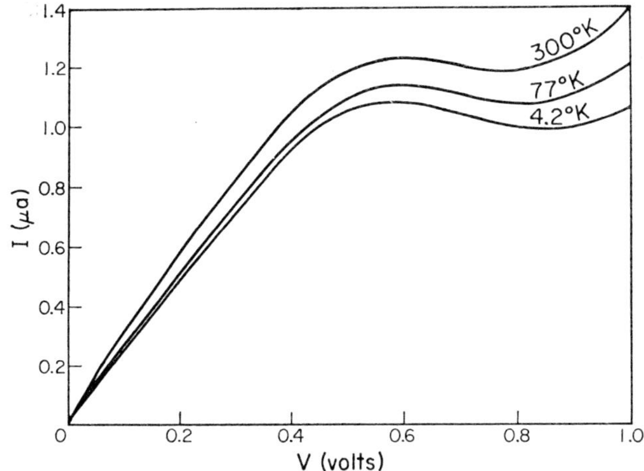

Fig. 13 J–V characteristic for Al–Al_2O_3–SnTe tunnel junction.

The qualitative features of the *I–V* characteristic can be explained using Fig. 14, which is an idealised energy diagram for the system for various voltage biases (the aluminium electrode is negatively-biased in all cases).

For voltages $eV < E_F$ (Fig. 14(b)) the electrons tunnel from the metal electrode into the empty states lying between the electrode Fermi level and the top of the valence band of the stannic telluride. The tunnel probability[(23)] ($\simeq \exp(-\Delta s\bar{\phi}^{1/2})$, where Δs and $\bar{\phi}$ are defined in Section 2.1(a)) for electrons at the Fermi level of the metal increases with increasing bias according to

$$\exp\left[-1{\cdot}025s(E_c + E_g + E_F - eV/2)^{1/2}\right] \qquad (19)$$

which means that the tunnel current into the stannic telluride valence band, I_v, increases with increasing bias.

When $E_F > eV > E_F + E_g$, Fig. 14(c), only electrons in

energy levels between AB in the electrode may tunnel into the stannic telluride. Electrons in levels in the electrode between A and the electrode Fermi level face the energy gap of the stannic telluride and hence cannot tunnel into the stannic telluride, because there are no available empty states there. The tunnel probability of electrons entering the stannic telluride just below the top of the valence band is given by

$$\exp\left[-1{\cdot}025s(E_g + E_c + eV/2)^{1/2}\right]$$

The tunnel probability now *decreases* with increasing voltage bias, which means I_v *decreases* for $E_F > eV > E_g + E_F$, resulting in the negative resistance region shown in Fig. 13.

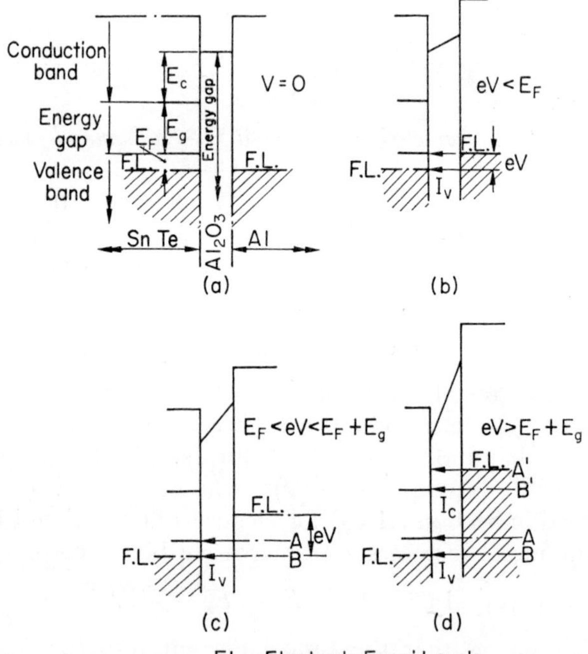

Fig. 14 Energy diagram for Al–Al_2O_3–SnTe tunnel junction for various voltage biases.

For $eV > E_F + E_g$ (Fig. 14(d)) the tunnel probability for the electrons tunnelling into the stannic telluride valence band is still given by equation (19). However, we now have an additional component of current, I_c, which is due to electrons in the electrode in levels between the Fermi level to the electrode and the bottom of the stannic telluride conduction band (i.e., between A' and B') tunneling into the *conduction* band of the electrode. The tunnel probability of an electron at the Fermi level is

$$\exp\left[-1{\cdot}025s(E_c + E_g + E_F - eV/2)^{1/2}\right]$$

which *increases* with increasing voltage bias; thus I_c *increases* with voltage bias. Since the total tunnel current is given by $I_c + I_v$, and, since I_c increases much more rapidly than I_v decreases, the tunnel current increases when $eV > E_F + E_g$.

In view of the above considerations and by inspection of Fig. 13, it can be seen that $E_F \simeq 0{\cdot}06$ eV and the energy gap of the stannic telluride is $E_g = 0{\cdot}30$ eV.

A quantitative account of the phenomena has been given by Chang *et al.*[(32)]

(e) TUNNELLING ANOMALIES

In recent years it has been observed that the tunnel conductance varies rapidly for voltage biases up to about 0·1 V. These observations have been attributed to the excitation by the tunnelling electrons of internal degrees of freedom in the barrier or the electrodes. Hence the tunnelling electrons may act as spectroscopic probes of the tunnel junction.

These deviations from the ideal tunnel I–V characteristics (Section 2.1(a)–(c)) are collectively referred to under the generic title of tunnelling anomalies. These anomalies can be divided into two broad categories: (i) zero-bias anomalies and (ii) finite-bias anomalies.

(*i*) *Zero-bias Anomalies*

This anomaly, which is manifest as a peak or dip in the junction conductance about zero bias, was first observed in III–V tunnel diodes.[33] The effect was first noted by Wyatt[34] (as a peak in the conductance) in metal-insulator-metal tunnel junctions. Wyatt studied tunnelling from bulk transition metals through the transition oxide to Al or Ag, and observed that the conductance peak had a logarithmic-voltage dependence and that the zero-bias conductance increased logarithmically with decreasing temperature. Shew & Rowell[38] have investigated the effect of magnetic field on the zero-bias anomaly. They have shown that at sufficiently high magnetic fields the zero-bias peak splits into two peaks positioned either side of zero-bias (see Fig. 15).

Applebaum[35,36] and Anderson[37] have explained the zero-bias conductance maxima in terms of excitation of

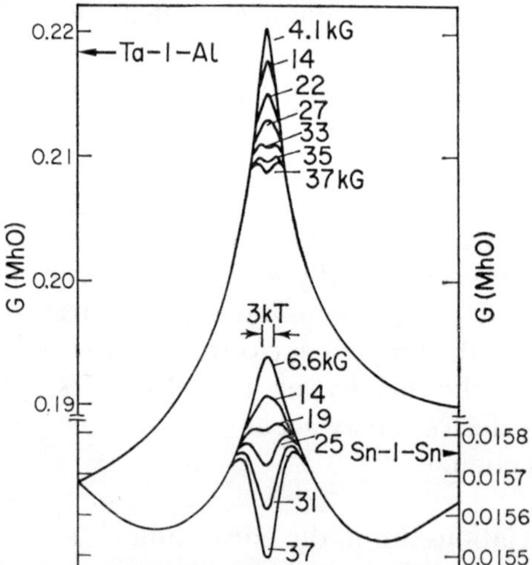

Fig. 15 Effect of magnetic field on zero-bias anomoly Ta-insulator-Sn Tunnel junctions. The temperature was 1·4°K and the voltage corresponding to 3kT(≃0·4 mV) is shown.

spin-flip scattering of electrons by non-interacting localised magnetic impurities existing at the metal-oxide interface of the transition-metal electrode. The tunnelling conductance for this process is shown schematically in Fig. 16, and may be regarded as the sum of three parts. The first term G_1 arises from electrons which either do not encounter an impurity or which are scattered without spin interaction. G_2 arises from current which is due to scattered electrons that flip the spin of the impurity. However, it is the third term G_3 that gives rise to the peak in the conductance, and is due to interference between transmitted electrons and electrons from the transition-metal

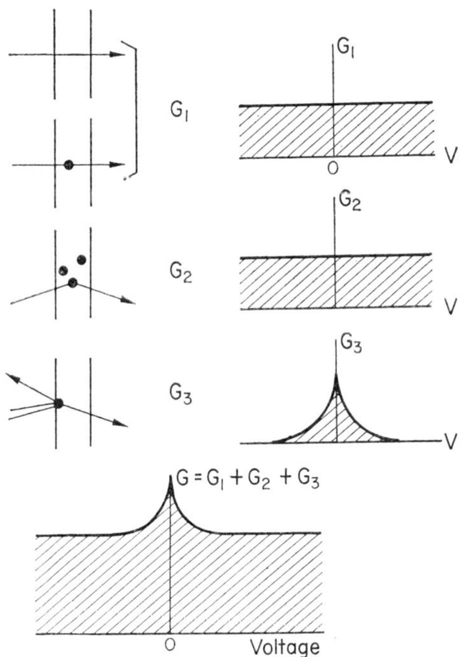

Fig. 16 Schematic illustration of the contributions to the conductance at zero temperature and in zero magnetic field in a junction containing localised paramagnetic impurities. The bottom diagram represents the total conductance.

electrode that are spin-flip scattered back into the latter electrode.

Applebaum[36] has shown that the splitting of the zero-bias conductance peak into two peaks about zero-bias in the presence of an applied magnetic field is due to a spin-flip transition incorporating an energy loss. This is because an applied field H splits the Zeeman levels of the

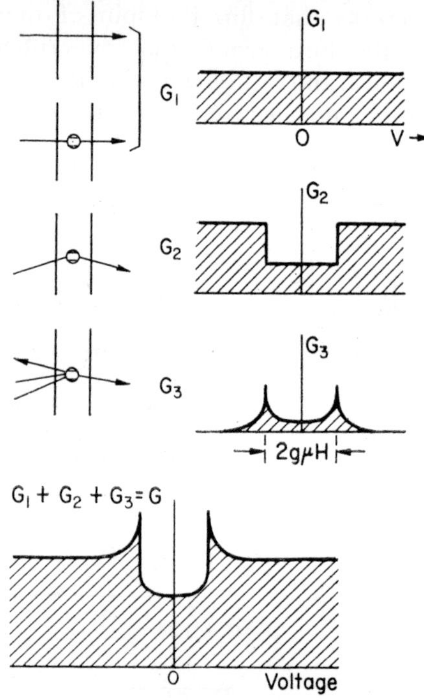

Fig. 17 Effect of magnetic field at 0°K on the three contribution to the conductance in a junction containing localised paramagnetic impurities. The sum of the various conductances is shown in the bottom diagram.

impurity by an amount $g\mu H_1$, and a tunnelling electron which flips the spin will have to lose this amount of energy. Thus the voltage bias across the junction must satisfy $V \geq g\mu H/e$ for the exchange to occur; hence a hole of voltage width ($2g\mu H/e$), centred on zero-bias, exists in the G_2 term (see Fig. 17). In order for the electrons to be spin-flip scattered back and interfere with the transmitted electrons, they also require energy at least equal to $g\mu H$; hence there will be *two* peaks in the G_3 conductance term, and these peaks will be positioned at $V = \pm g\mu H/e$. Clearly then these tunnelling processes manifest in a

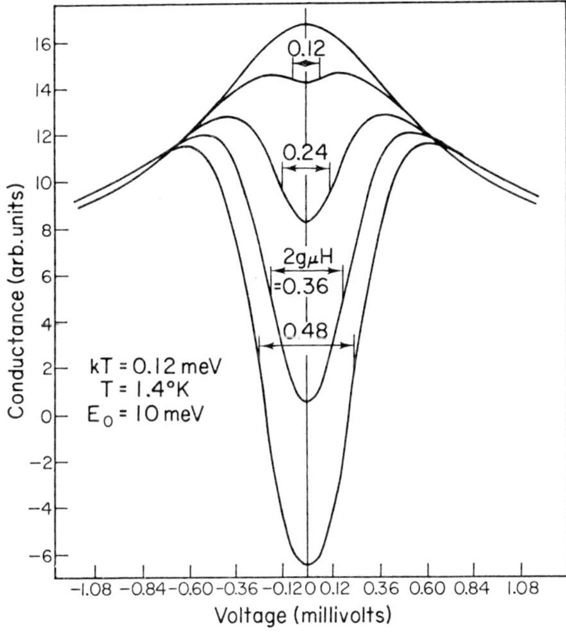

Fig. 18 Appelbaum's calculations of the effect of a magnetic field on the zero-bias tunnelling anomaly (at $T = 1{\cdot}4°K$). Four values of $2g\mu H$ are shown.

qualitative manner the salient features of the experimental observations of Shen and Rowell.[38] In actual fact, a quantitative analysis[36] yields curves (Fig. 18) which are remarkably similar in appearance to the experimental data, and leave little doubt about the nature of the mechanism involved.

(ii) Finite-bias Anomalies

These anomalies involve an energy exchange between the tunnelling electron and some internal mode of the insulator, such as phonon excitations in the oxide and the electrodes,[(25)] and excitations of the vibrations of molecules trapped in the insulator.[(39,40)]

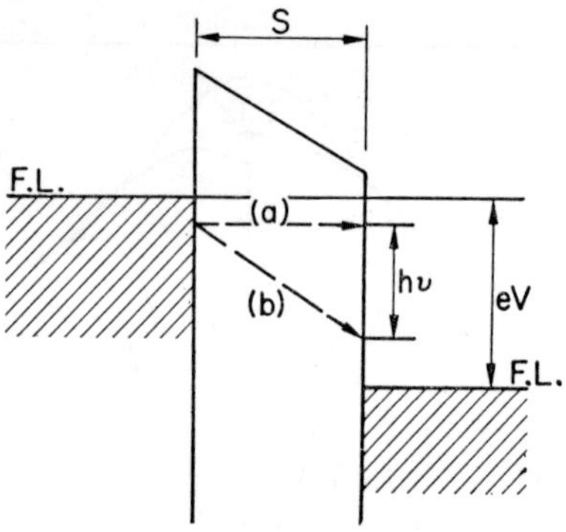

Fig. 19 (a) Schematic energy-level diagram for tunnelling between normal metals. Path (a) represents elastic and (b) inelastic tunnelling channels. The inelastic process can occur only if $eV > h\nu$.

Figure 19 is an energy diagram of a tunnel junction depicting the energy-loss mechanism. The upper path indicates an electron making a constant energy (elastic) transfer between the electrodes, and the lower path an inelastic transfer. It will be clear from this diagram that when $eV < h\nu$, an electron from the negatively-biased electrode cannot excite the mode (of frequency ν), for it would be required to enter the positively-biased electrode at an energy level positioned below the Fermi level, and all such levels are filled according to the Pauli Exclusion Principle. Thus the current for $V < h\nu/e$ is due only to elastic tunnelling. When $eV > h\nu$, both elastic and inelastic tunnelling become possible. What this means experimentally is that structure will be observed in the

J–V characteristic as the voltage sweeps through a value $h\nu/e$. In principle, it is possible to excite all the internal modes of the junction, which means that structure in the *J–V* characteristic will be apparent at more than one voltage. In practice, it is necessary to measure the first or second derivative of the current with respect to voltage in order to observe the excitation, since the magnitude of the current resulting from inelastic processes is small compared to that of the elastic process. Techniques for measuring first and second derivatives are reported in the literature.[41] Figure 20 shows the first and second derivative for a Pb–I–Pb junction.[25] It is believed that the second derivative peaks between 25 and 60 mV are due to the excitation of the oxygen atom of the lead oxide. The peak at 17 mV is due possibly to a lead atom frequency. The peaks at 5 and 9 meV are characteristic phonon energies for bulk lead.

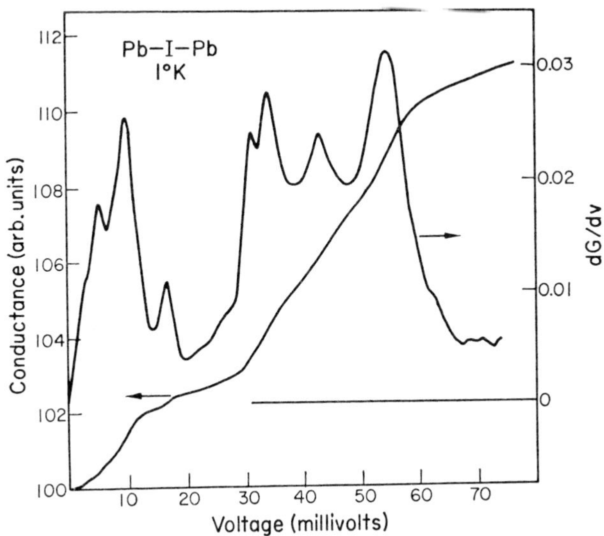

Fig. 20 Conductance (dI/dV) *and derivative of conductance* (d^2I/dV^2) *for a Pb-insulator-Pb junction at 1°K.*

2.2 THERMIONIC EMISSION

When the potential barrier is too thick to permit tunnelling to occur, or at sufficiently high temperature, the current flowing through the insulator is limited principally by the rate at which electrons are thermally-excited *over* the interfacial potential barrier into the insulator conduction band. Thus, in order to determine the current flowing in the system we set the lower limit in the integral over dE_x in equation (6) to be equal to the interfacial barrier height ϕ_0, and the transmission function $P(E_x)$ to unity. Equation (6) then integrates to

$$J = (4\pi mek^2T^2/h^3) \exp(-\phi_0/kT) \equiv AT^2 \exp(-\phi_0 kT) \tag{19}$$

which is just the Richardson (saturated) thermionic emission equation, if ϕ_0 is independent of the voltage bias. (Note that A takes the value 120 when J is expressed in A cm^{-2}). In point of fact, because of attendant image forces due to electrons polarising the surface of the electrode from which they were ejected, ϕ_0 varies with applied field. Thus the current does not saturate, but is a function of the applied field.

The potential energy of the electron due to the image force is

$$\phi_{im} = -e^2/16\pi e_0 K^* x \tag{20}$$

where x is the distance of the electron from the electrode surface. The dielectric constant K^* in equation (20) is the *high-frequency* constant, since in the course of emission from the cathode the escaping electrons spend only an extremely short time in the immediate vicinity of the surface.

(a) NEUTRAL CONTACT

The potential step (with respect to the Fermi level) at a neutral barrier (Section 1.4(c)) with attendant image potential as a function of the distance x from the interface is given by (see Fig. 21)

$$\phi(x) = \phi_0 + \phi_{im} = \phi_0 - e^2/16\pi\varepsilon_0 K^* x \qquad (21)$$

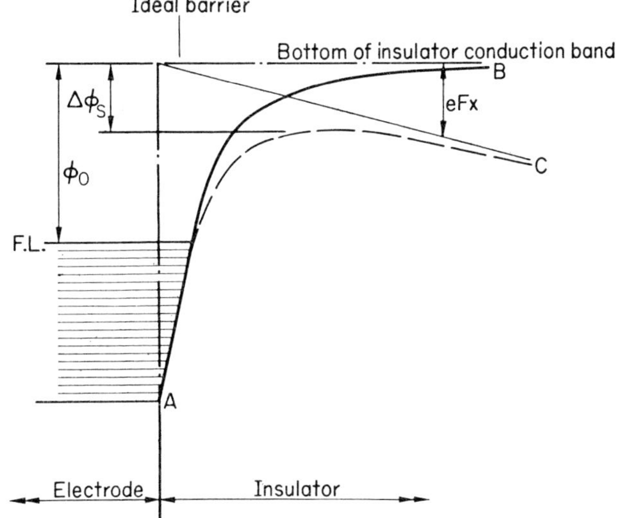

Fig. 21 Schottky effect at a neutral contact.

If a uniform field exists at the surface of the cathode, $\phi(x)$ is given by

$$\phi(x) = \phi_0 - e^2/16\pi K^*\varepsilon_0 x - eFx.$$

This equation has a maximum at $x_m = (e/16\pi K^*\varepsilon_0 F)^{1/2}$. The attenuation $\Delta\phi_s (= \phi_0 - \phi(x_m))$ of the barrier height due to the interaction of the applied field with the image potential, which is the Schottky effect, is therefore given by (see Fig. 21):

$$\Delta\phi_s = (e^3/4\pi K^*\varepsilon_0)^{1/2} F^{1/2} \equiv \beta_s F^{1/2} \qquad (22)$$

Substituting $\phi(x_m) = \phi_0 - \Delta\phi_s$ for ϕ in equation (19) we obtain

$$J = AT^2 \exp(-\phi_0/kT) \exp(\beta_s F^{1/2}/kT) \qquad (23)$$

which is just the Richardson–Schottky equation.[42] Equation (23) was first applied successfully to metal-vacuum interfaces. The effect in insulators at a neutral contact appears to have been first observed by Emptage and Tantraporn,[43] who reported a log I versus $F^{1/2}$ relationship in their samples; since then there have been many other reported similar observations.

When sufficiently high fields exist at the interface, tunnelling through the potential barrier can dominate the conduction process, and the relevant I–V characteristic is given by equation (18) suitably modified to include image forces.[23,24] This mechanism requires that the thickness of the barrier, measured at the negatively biased electrode Fermi energy, be less than about 50 Å for observable current flow. Thus the onset of field emission at a neutral contact occurs at a field given by

$$F \simeq \phi_0 \,.\, 10^8/50 = 2 \times 10^6 \phi_0 \;\; \text{Vcm}^{-1}.$$

(b) BLOCKING CONTACT

The potential energy of a blocking contact of the Schottky-barrier type (Section 1.4(b)) with attendant image potential is,[44]

$$\phi(x) = \phi_0 - (N_d e^2/K\varepsilon_0)(\lambda x - x^2/2) - e^2/16\pi K^* \varepsilon_0 x \tag{24}$$

This barrier has a maximum at $x_m \simeq 1/4(\pi N_d \lambda)^{1/2}$. In this case the image force lowers the potential barrier by an amount

$$\Delta\phi = \phi_0 - \phi(x_m) = \left[\frac{(\text{eV} + \psi_m - \psi_i)e^7 N_d}{2(8\pi)^2(\varepsilon_0 K^*)^3}\right]^{1/4}$$

Substituting $\phi(x_m)(=\phi_0 - \Delta\phi)$ for ϕ_0 in equation (19) yields, when ϕ_0, ψ_m, ψ_i and kT are expressed in electron volts and N_d in cm^{-3}:

$$J = 120T^2 \exp \left\{ -\frac{\phi_0 - 8{\cdot}26 \times 10^{-6} \, [N_d(V + \psi_m - \psi_i)/K^{*3}]^{1/4}}{kT} \right\} \text{A cm}^{-2} \tag{25}$$

As at the neutral contact, tunnelling will occur within the barrier when the field at the interface is sufficiently high to reduce the width of the barrier measured at the electrode Fermi level to about 50 Å or so. For $V > \phi_0$, the barrier at the interface can be considered approximately triangular, so we may use the Fowler–Nordheim expression, equation (18), to estimate the current tunnelling through the contact, which is given by:[(2)]

$$J = \frac{1{\cdot}22 \times 10^{-11} N_d(V + \psi_m - \psi_i)}{\phi_0 K^{*3}} \times \exp\left[-\frac{3{\cdot}65 \times 10^{10} (K^* \phi_0{}^3)^{1/2}}{N_d(V + \psi_m - \psi_i)^{1/2}} \right] \tag{26}$$

We are not aware of any incontrovertible experimental evidence for the processes in thin-film insulators characteristic of the types given by equations (25) and (26). However, the tunnelling process equation (26) is probably the dominent mechanism in the electrode limited range described in Section (4.2). Actually, Schottky barriers in thin films are much more easily investigated using a.c. techniques.[(45)]

3. Bulk-limited Conduction

3.1 SPACE CHARGE LIMITED CURRENTS

(a) THEORY

If the electrodes make ohmic contacts to the insulator, of the type shown in the energy diagram of Figs. 3(c) and 22(a), the conduction process is space-charge limited (SCL). It will be recalled that in order to form the ohmic contact shown in Figs. 3(c) and 22(a), a *negative* space-charge region exists inside the surfaces of the insulator, and an equal and opposite (positive) surface charge exists on the electrodes. The application of a voltage bias to the

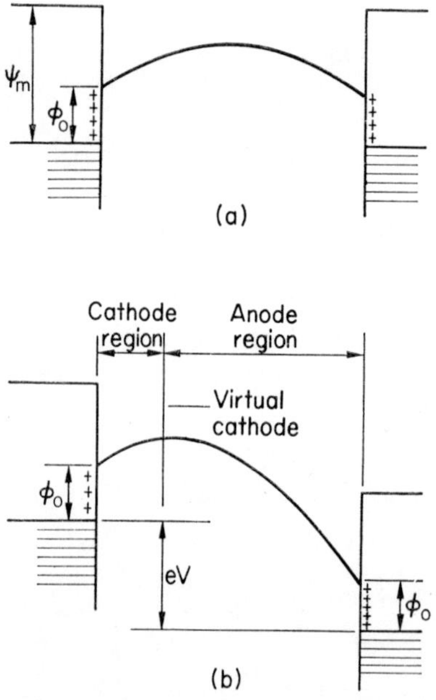

Fig. 22 Insulator containing ohmic contacts (a) under zero bias (cf. Fig. 3(c)) (b) with voltage bias.

system results in an increase in and redistribution of the space charge in the insulator and an increase and decrease respectively in the positive surface charge existing on the anode and the cathode (see Fig. 22(b)). The positive charge on the cathode is just equal to the negative space charge existing in the cathode region, that is the region between the cathode-insulator interface and the virtual cathode (the plane at which the field in the cathode is zero). Similarly, the positive charge on the anode is just equal to the space charge existing throughout the anode region (that is the region between the virtual cathode and the anode-insulator interface). The free component of the space charge existing in the insulator conducts the current; hence the term space charge limited current.

The cathode region can be considered to be a charge reservoir, that supplies charge to the anode region as demanded by voltage bias conditions. Thus, provided the cathode region exists, that is *positive* charge exists on the cathode, the conduction process will be SCL. Mott and Gurney[46] made the first theoretical study of SCL currents in insulators. They confined their studies to a trap-free insulator, for which they obtain the following relationship:

$$J = (9\mu K\varepsilon_0/8s^3)V^2 \tag{27}$$

Equation (27) predicts that SCL current is proportional to V^2 and inversely proportional to the s^3. Both of these predictions have since been confirmed by experiment. However, equation (27) predicts much higher currents than are observed in practice, and also that the current is temperature insensitive, which is also contrary to observation. These deviations from the simple trap-free theory are readily accounted for when a more realistic insulator, that is, one containing traps, is considered.

The first study of SCL currents in defect insulators was made by Rose.[47] He points out that when the insulator contains traps, a large fraction of the injected space-charge condenses into the traps. Thus, only a fraction of the charge drawn into the insulator by the applied voltage is available to conduct current, whereas in a trap-free insulator all the space charge is available to participate in the conduction process. The ratio θ of free-to-trapped charge in an insulator containing *shallow* traps is[47]

$$\theta = (N_c/N_t) \exp(-E_t/kT) \tag{28}$$

where N_t is the trap density and E_t is the depth of the trap below the bottom of the conduction band. Assuming $N_t = 10^{19}\ \text{cm}^{-3}$ and $E_t = 0{\cdot}25$ eV, and taking $N_c = 10^{-19}\ \text{cm}^{-3}$ at room temperature, we have $\theta < 10^{-5}$. To incorporate the effect of shallow traps into the theory all that is required is to multiply the right side of equation (27) by θ:

$$J = (9\mu K\varepsilon_0\theta/8s^3)V^2 \tag{29}$$

It will be clear, since θ is independent of V, $J \propto V^2$ as in the trap-free case. Also, we have seen that θ is a very small quantity and temperature dependent; thus the inclusion of shallow traps in the insulator brings the theory into line with experimental observation.

Only when the injected *free-carrier* density, n_i, exceeds the volume-generated free-carrier density, n_0, will space-charge effects be observed;[48,49] when $n_0 > n_i$ the volume (ohmic) conductivity will predominate.

The voltage, V_x, at which the transition from ohmic to SCL conduction occurs is[48,49]

$$V_x = en_0s^2/\theta K\varepsilon \tag{30}$$

When sufficient charge is injected into the insulator, the traps will be saturated (trap-filled limit, TFL). Thus, any additional charge injected into the insulator exists as free charge in the conduction band and contributes *in toto* to the current. Beyond the TFL, the insulator behaves as if it were trap-free with the result that the J–V characteristic is given by equation (27) rather than equation (29). Hence as V just exceeds V_{TFL}, the current rises rapidly by an amount θ^{-1}. The voltage at which the TFL occurs is given by

$$V_{\mathrm{TFL}} = eN_t s^2/2K\varepsilon_0 \tag{31}$$

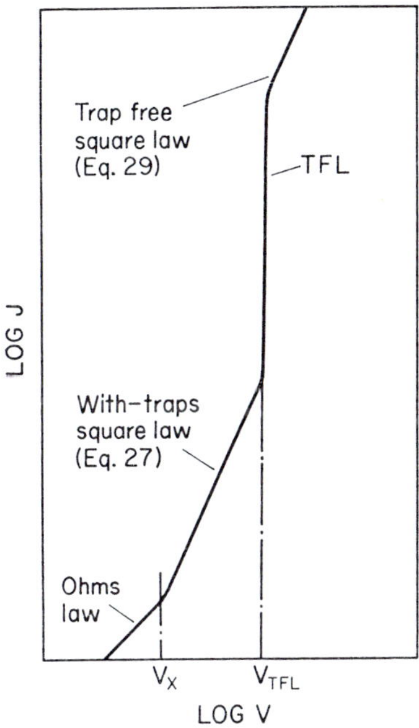

Fig. 23 S.C.L. J–V characteristic for an insulator containing shallow traps.

In Fig. 23 we have illustrated schematically a typical J–V characteristic for an insulator having a shallow discrete trapping level. At the lower voltages ($V < V_x$), when the bulk-generated current exceeds the SCL current, the characteristic is ohmic. In the voltage range $V_x < V < V_{\mathrm{TFL}}$ the SCL current predominates and J is proportional

to V^2 (cf. equation (29)). When $V = V_{TFL}$, sufficient charge has been injected into the insulator just to fill the traps. Hence, as V just exceeds V_{TFL}, the current rises rapidly such that for $V > V_{TFL}$ the J–V characteristic obeys the trap-free law, equation (27). Clearly, from the structure exhibited in the characteristic, much information about traps in the insulator can be deduced from the experimental data.

Rose[47] has treated the case of SCL conduction in the presence of a *distribution* of trap levels that decreases exponentially in density with increasing energy below the conduction band; that is, $N_t = A \exp(-E/kT_c)$, where E is the energy measured from the bottom of the conduction band, and T_c is a characteristic temperature greater than the temperature at which the currents are measured. This trap distribution is typical of that which can be expected to exist in an amorphous solid. For this case

$$J \propto V^{(1+T_c/T)} \tag{31a}$$

which means that, since $T_c > T$, the current increases more rapidly with applied voltage than in the case of the trap-free or discrete trap-level cases

(b) EXPERIMENTAL

Budinas *et al.*[50] have studied space-charge limited current flow in vitreous antimony tripulfide (Sb_3S_3) dielectric films. Bismuth electrodes were used to provide ohmic contracts to the dielectric. The J–V characteristics for a sample of thickness 11·1 μm measured at $T = 290°$K are shown in Fig. 24. At the lower voltages ($V < 100$ V) the I–V characteristic is ohmic, and immediately beyond this region the current shows a quadratic dependence on voltage. At about 600 V there is a transition from the quadratic dependence of the current on voltage to that of a steeply-rising current.

The voltage, V_x, at which the transition from the linear to the quadratic dependence of current on voltage occurred, was found to be proportional to s^2 (cf. equation (30)) as shown by curve 1 in Fig. 25. Also, in the region where $J \propto V^2$ it was found that $J \propto s^{-3}$ (cf. equation (29)) as shown by curve 1 in Fig. 26. It was therefore concluded that beyond V_x the current was SCL in the presence of a shallow discrete trapping level. The SCL nature of the conduction process was confirmed by the resulting J–V characteristics of the sample under white-light illumination, since for $V > V_x$ the decrease in the total current (photocurrent plus dark current) to dark current is observed,[51] as shown in Fig. 25.

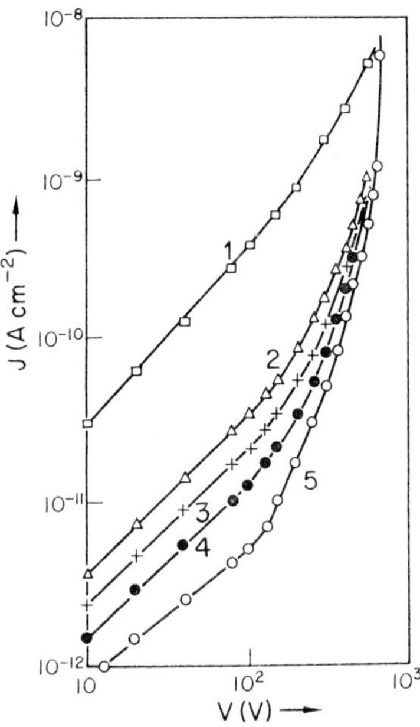

Fig. 24 J–V characteristics[50] for Sb_2S_3 film at $T = 290°K$ and at different light intensities in terms of an arbitrary light unit B_0. Curve 1, intensity $= B_0$; Curve 2, intensity $= 10^{-2}B_0$; Curve 3, intensity $= 3{\cdot}5 \times 10^{-3}B_0$; Curve 4, intensity $= 10^{-3}B_0$; Curve 5, dark J–V characteristic. Parameters of sample junction area $= 9{\cdot}5 \times 10^{-3}$ cm^{-2}, $S = 11{\cdot}1 \mu m$.

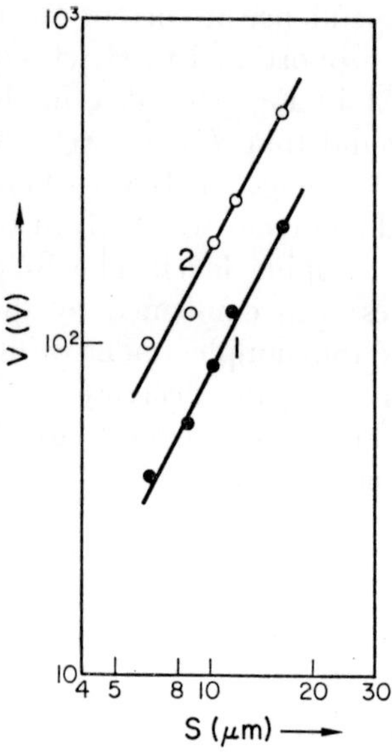

Fig. 25 Dependence of V_X and V_{TLF} on Sb_2S_3 fim thickness.(50) *Curve 1 illustrates V_X versus S and curve 2, V_{TLF} versus s.*

The voltage for the onset of the rapid increase in the J–V characteristic beyond the $J \propto V^2$ region was identified as V_{TFL}, since this voltage was found to be proportional to s^2 (cf. equation (31)), as shown by curve 2 in Fig. 25. Also, the current at the transition voltage was found to be proportional to s (see curve 2 in Fig. 26)—this relationship is obtained by substituting the expression for V_{TFL} in equation (29). From these observations Budinas obtained the following values for the parameters of their films assuming $K = 13$, $\mu = 1$ cm^2 V^{-1} s^{-1}, $N_t = 3{\cdot}5 \times 10^{15}$ cm^{-3}, $E_t = 0{\cdot}78$ eV and $\theta = 2{\cdot}6 \times 10^{10}$. Good confirmation for the two former parameters, N_t and E_t was obtained by plotting θ against T^{-1} (see equation (35)); this plot, which was linear, yielded values of $N_t = 4 \times 10^{15}$ cm^{-3} and $E_t \simeq 0{\cdot}76$ eV.

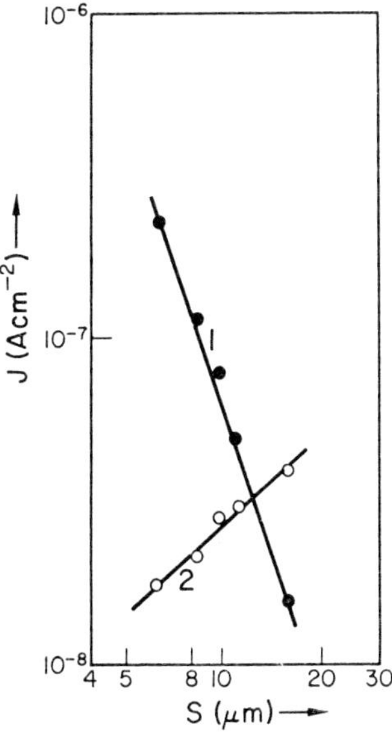

Fig. 26 Dependence of current density on Sb_2S_3 film thickness.[50] *Curve 1 illustrates the dependence of the current in the SCL region (before TFL) on film thickness at $V = 350$ V. Curve 2 illustrates the dependence of the current at $V = V_{TLF}$ on insulator thickness.*

3.2 POOLE–FRENKEL EFFECT

When an electric field interacts with the coulombic potential barrier of a donor centre or trap, the height of the barrier is lowered, as shown in Fig. 27. This process is the bulk analogue of the Schottky effect at an interfacial barrier. Since the potential energy of an electron in a Coulombic field, $-e^2/4\pi\varepsilon_0 K^* x$, is four times that due to image force effects, the Poole–Frenkel attenuation of a Coulombic barrier, ϕ_{PF}, in a uniform electric field is twice that due to the Schottky effect at a *neutral* barrier:

$$\Delta\phi_{PF} = (e^3/\pi\varepsilon_0\, K^*)^{1/2} F^{1/2} \equiv \beta_{PF} F^{1/2} \qquad (32)$$

Frenkel[52,53] suggested that the ionisation potential, E_g, of the atoms in a solid is lowered an amount given by equation (32) in the presence of a uniform field. Thus the conductivity is field-dependent and of the form

$$\sigma = \sigma_0 \exp\,(\beta_{PF}F^{1/2}/2kT) \tag{33}$$

where σ_0 $(=e\mu N_c \exp\,(-E_g/2kT))$ is the low-field conductivity. Equation (33) may be written in the form

$$J = J_0 \exp\,(\beta_{PF}F^{1/2}/2kT) \tag{34}$$

where J_0 $(=\sigma_0 F)$ is the low-field current density

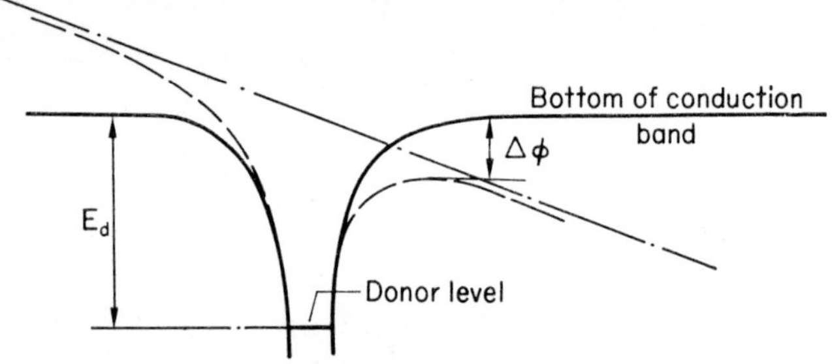

Fig. 27 Poole–Frenkel effect at a donor centre.

It is interesting to note that although $\Delta\phi_{PF} = 2\Delta\phi_s$ the coefficient of $F^{1/2}$ in the exponential is the same for both the Richardson–Schottky (at a neutral contact, see equation (23)) and Poole–Frenkel J–F characteristics (i.e. $\beta_{PF}/2 = \beta_s$). Mead[54] has suggested, however, that since traps abound in an insulator and that a trap having a Coulombic-type barrier would experience the Poole–Frenkel effect at high fields, thereby increasing the probability of escape of an electron immobilised therein, the current density in thin-film insulators containing *shallow* traps is given by:

$$I = J_0 \exp\,(\beta_{PF}F^{1/2}/kT) \tag{35}$$

Note in this case that the coefficient of $F^{1/2}$ is *twice* that in equation (34). Mead[54] also first reported field-dependent conductivity apparently of the form given by equation (35), and for this reason equation (35) is usually the form of Poole–Frenkel equation associated with *thin-film* insulators rather than that given by equation (34).

Several investigators have observed a field-dependent conductivity of the form given by equation (35) in what is apparently *bulk-limited* conduction in Ta_2O_5 and SiO films.[55–57] Johanson[56] has noted, however, that the coefficient of $F^{1/2}/kT$ is compatible with the Schottky effect rather than the Poole–Frenkel effect. On closer examination of the results of Mead[54] and of Hirose and Wada,[55] although they claim Poole–Frenkel emission, it is apparent that their experimental β are actually compatible with Schottky emission.[1] These authors based their conclusions on dielectric values which are typically four times too high for the high-frequency dielectric constants of the materials used. Hartman *et al.*[57] conclude that neither the Richardson–Schottky equation (24), because conductivity is bulk-limited, nor the Poole–Frenkel equation (35), because of the incompatibility of the experimental and theoretical β, can adequately explain their results.

In actual fact, in films of tantalum oxide (Ta_2O_5) and silicon oxide, because of their wide energy gaps, in order to have any detectable current, even in the absence of traps, it is necessary to have donor or acceptor centres within the insulator to supply the necessary carriers.[1] Furthermore, if it is assumed that the insulator contains donor centres which lie below the Fermi level (an assumption supported by the fact that the conductivity of the films continues to increase with increasing temperature above room temperature) and shallow *neutral* traps (see Fig. 28), the bulk *I–V* characteristic of the film is given by[1]

$$J = J_0 \exp (\beta_{PF} F^{1/2}/2kT) \tag{36}$$

where $J_0 = e\mu N_c (N_d/N_{ts})^{1/2} F \exp [-(E_d + E_{ts})/2kT]$ (37)

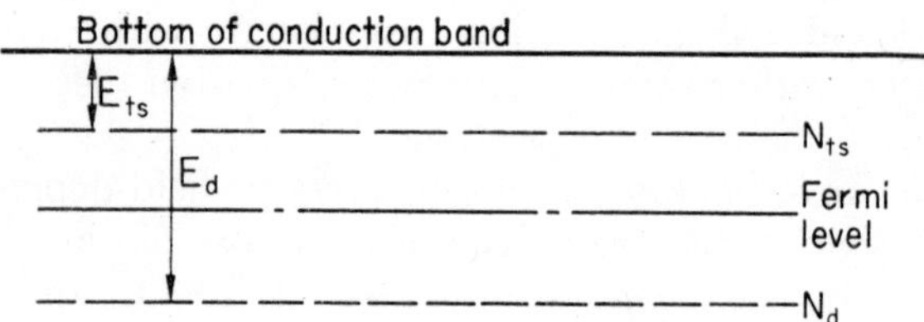

Fig. 28 Energy diagram of insulator film used in deriving (37).

Thus, in this case, the coefficient of $F^{1/2}/kT$ is $\beta_{PF}/2(=\beta_S)$ *even though the conductivity is not electrode-limited,* which explains the anomalous experimental results. It should be noted that equation (37) will hold even if there exists a range of trap and donor levels and one of the trap levels is distinctly more predominant than others of their species.

Stuart[58,59] has shown that his bulk-limited I–V characteristics, which are observed to be linear on a ln I versus $V^{1/2}$ plot, can be explained in terms of equation (36). In Table 2 we have transcribed his calculation of the ratios of the experimentally determined $\beta(\beta_{exp})$ and the theoretical $\beta(\beta_{theor})$ assuming equation (36) as the relevant I–V characteristic. The agreement between β_{exp} and β_{theor} is seen to be excellent over the entire thickness range, which covers almost *four* decades.

Table 2. Stuart's (1967(a,b)) Results for $\beta_{exp}/\beta_{theor}$ Obtained from SiO *Films*

s (microns)	0·043	0·12	0·32	0·64	1·2	3·3	7·0	13·7
$\beta_{exp}/\beta_{theor}$	1·06	1·07	1·09	1·07	1·04	1·05	1·03	1·00

It will be apparent from equation (33) that a plot of ln J versus T^{-1} at constant voltage should yield a straight line of slope $(\beta_{PF}F^{1/2} - E_d - E_{ts})/k$. Such plots are normally observed in wideband dielectrics[54,57–60] at sufficiently high temperature, indicating that the conduction is by free electrons in the conduction band. Since β_{PF} can be determined from isothermal measurements, this method provides a means of determining $(E_d + E_t)$. (See equation (37).)

4. Translational Conduction Processes

4.1 BULK LIMITED TO ELECTRODE LIMITED PROCESS

When an insulator with ohmic contacts on its surface, as shown in Fig. 3(c), is subjected to sufficiently high voltages there occurs a bulk-limited (SCLC, see Section 3.1) to electrode limited (Schottky-type) transition in the conduction process.[61]

Consider Fig. 29, which illustrates the energy diagram for an insulator containing ohmic contacts under voltage bias. It will be recalled (Section 3.1(a)) that while a positive charge exists on the cathode the conduction process is bulk (SCL) limited. Thus the I–V characteristic of an insulator will be given by equations (27), (29) or (31(a)) depending on the circumstances. Since the positive charge on the cathode decreases with increasing voltage bias (Fig. 29), at a sufficiently high voltage bias, V_T, this charge is reduced to zero, and for voltages greater than V_T a negative charge appears on the cathode. Thus, the current at this stage ceases to be space-charge limited, and is now limited by the rate at which carriers can flow across the cathode barrier from the electrode into the insulator; in other words the current is now electrode limited. *Hence the presence of negative charge on the cathode is a major distinction between the space-charge-limited and emission-limited regions.* In the space-charge limited region there was *positive* charge on the cathode which was depleted only at the onset of emission-limited current.

The current required to first go emission limited is given by

Richardson equation for thermionic emission (see equation (23)).

$$J_T \simeq AT^2 \exp(-\phi_0/kT) \quad (38)$$

where ϕ_0 is the barrier height at the cathode-insulator

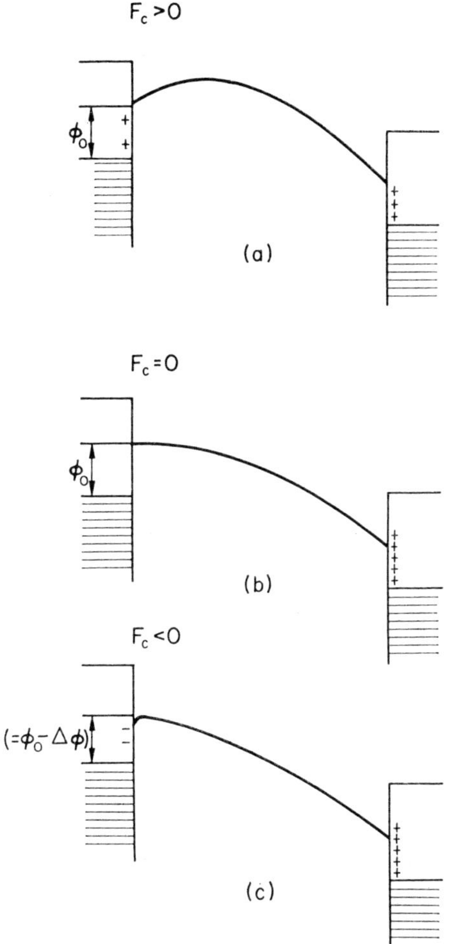

Fig. 29 Insulator with ohmic contacts under following voltage biases: (a) $V < V_T$ (b) $V = V_T$ and (c) $V > V_T$.

interface. If it is assumed that the insulator contains only shallow traps and that $V_T < F_{TFL}$ (see equation (31)), then this current is also given by

$$J_\sigma = \frac{9}{8} \frac{\mu\varepsilon\theta}{s^3} V_T^2 \tag{39}$$

From equations (38) and (39) we find that

$$V_T = T \left(\frac{8s^3 A}{9\mu\varepsilon\theta} \right)^{1/2} \exp \left(-\frac{\phi_0}{2kT} \right) \tag{40}$$

Thus the onset of emission-limited current in an insulator containing shallow traps is shifted towards higher voltages for smaller μ, θ, or ϕ_0.

If the height of the cathode barrier ϕ_0 were independent of the voltage bias, then the emission-limited current would saturate to a *value* given by the Richardson formula, equation (38). However, since negative charges accumulate on the cathode surface as V exceeds V_T, this means that a negative field now exists at the interface. Thus the cathode barrier is lowered an amount $\Delta\phi$ by the interaction of the image force with the field at the interface (Schottky effect—see Section 2.2a) as follows:

$$\Delta\phi = \beta_s F_c^{1/2}$$

where F_c is the field at the cathode-insulator interface. Thus for $V > V_T$ the current flowing through the system is given by (cf. equation (23)).

$$J = AT^2 \exp(-\phi_0/kT) \exp(\beta_s F_c^{1/2}/kT) \tag{41}$$

Because of the space charge in the insulator, F_c is not linearly dependent on V; in fact, F_c cannot be determined as an analytical function of V. Thus, one has to resort to numerical methods to obtain F_c in terms of V and the system parameters (for further details, the reader is referred to the literature.[61])

Some numerical evaluations for various barrier heights

and thicknesses[61] are shown in Fig. 30. A dielectric constant $K = 4$, $T = 300°K$, and $A = 120$ A/cm are assumed. The curves begin at the onset of ELC and continue until they asymptote to the sloping straight line. The straight line has the same form as equation (23), the *J–V* characteristic for ELC at a neutral contact. Hence as the voltages increase beyond V_T the effects of space charge

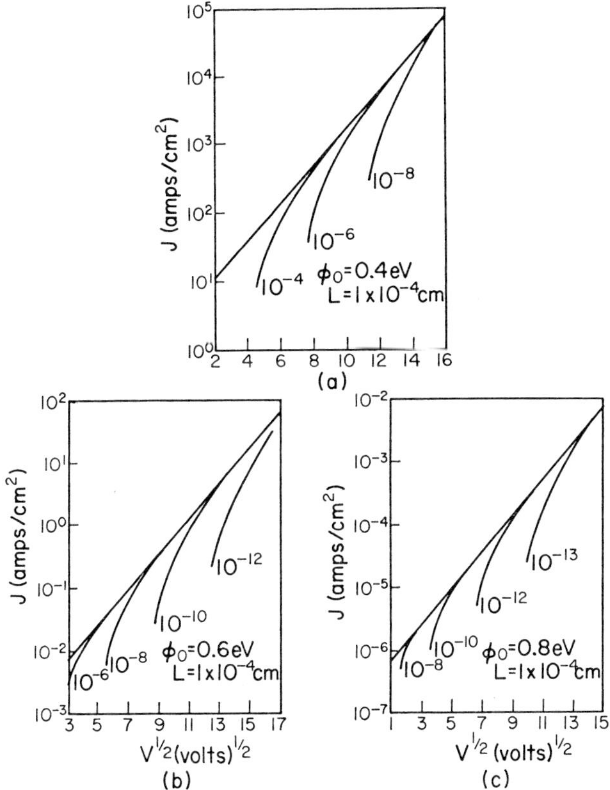

Fig. 30 J–V characteristics for emission limited current. J is given in A/cm², V = volts. K = 4 in all cases. $\mu\theta_0$ is shown for each curve where μ = mobility in cm²/V–s and θ_0 = ratio of free to trapped charge at zero bias. The curves begin at the onset of ELC.

are diminished, until at some sufficiently high voltage they are completely abrogated. What this means in effect is that the field throughout the insulator is essentially uniform and equal to V/L.

Schug *et al.*[62] have analysed their data obtained from Mylar films on emission-limited theory given above.

4.2 ELECTRODE-LIMITED TO BULK-LIMITED PROCESS

Consider an insulator containing a high density ($\simeq 10^{19}$ cm^{-3}) of donors and deep traps (see Fig. 31). If metal electrodes of work function $\psi_m > \chi + E_c - E_{td}$ are

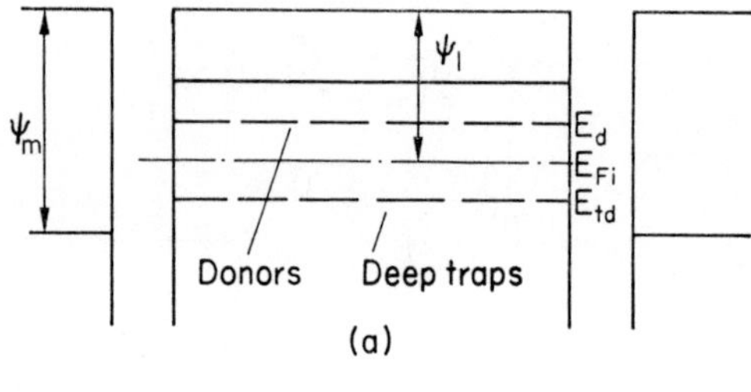

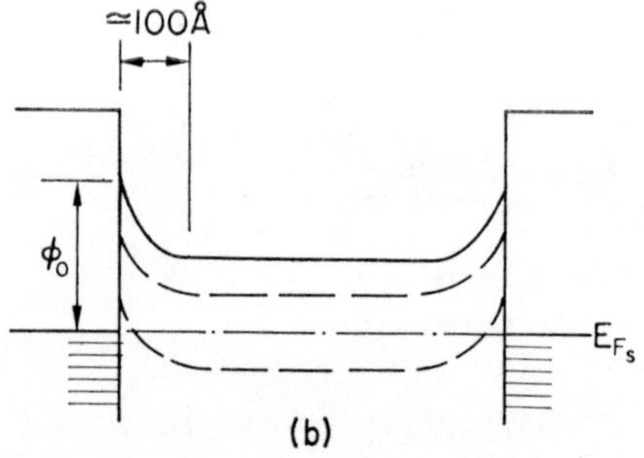

Fig. 31 Energy of metal-insulator-metal system of the type required to observe electode- to bulk-limited conduction transition: (a) *before electrodes are attached* (b) *after electrodes are attached.*

applied to the surface a blocking contact having a very narrow depletion region is formed (see Table 1) as shown in Fig. 31(b). Also, the "interior" of the insulator is of relatively high resistivity (see equation (33)) since essentially all the electrons given up by the donors to the conduction band have been captured and immobilised by the deep traps. In this system, during the initial application of voltage, field emission of electrons from the cathode into the conduction band of the insulator occurs and the current rises rapidly with applied voltage (see equation (26)). Since the contact resistance is much higher than that of the bulk, it follows that the *I–V* characteristic will be virtually *thickness-independent* (because essentially all of the applied voltage appears across the contact and very little across the bulk) and very steep (see chain dotted line in Fig. 32).

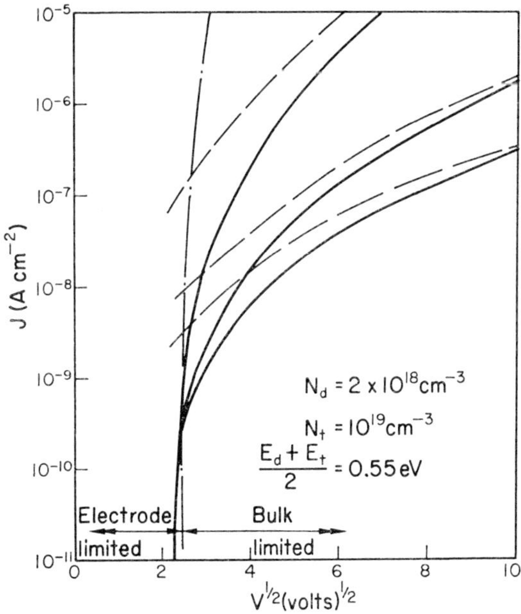

Fig. 32 Theoretical curves showing electrode- to bulk-limited characteristic.

The electrode-limited process cannot continue indefinitely, however, because the bulk resistance (see equation (35)) decreases much slower with increasing voltage than does the contact resistance (compare dotted and chain-dotted lines in Fig. 32). Thus, at some voltage V_T, the transition voltage, the contact resistance falls to a value equal to that of the bulk, and this occurs when the applied voltage, V, is shared equally between the contact and the bulk. Thereafter, practically all the voltage in excess of V_T falls across the bulk and the remaining fraction across the

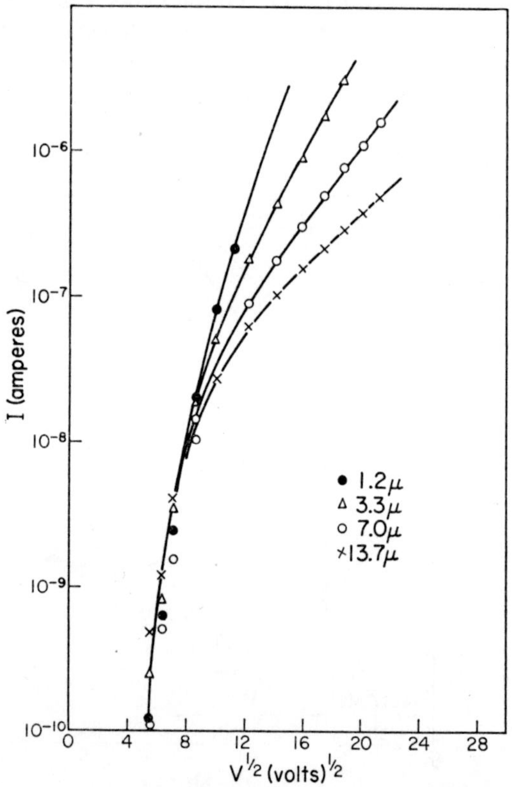

Fig. 33 Electrode-limited to bulk-limited conduction in Al–SiO–Al films. Electrode area 0·1 cm². (After Stuart, 1967b.)

barrier, just sufficient to ensure current continuity throughout the system. Hence, for voltages in excess of V_T, the current ceases to rise as rapidly as for $V < V_T$, since it is controlled by bulk processes, and will be *thickness dependent.*

In Fig. 32, the theoretical I–V characteristics for three thicknesses of dielectric, ($S = 1$, 3 and 10μ) have been plotted, using equations (26) and (35) and the parameters $N_d = 10^{18}$ cm^{-3}, $N_t = 10^{19}$ cm^{-3}, $K^* = 3$, $\mu = 10$ cm^2 (V s)$^{-1}$, $(E_t + E_d)/2 = 0{\cdot}55$ eV and $\phi = 1{\cdot}5$ eV. The dotted and chain-dotted lines represent the bulk and contact characteristics given by equations (35) and (26), respectively. The full lines represent the combined effect of the contact and bulk; that is, the actual characteristic of the junction. The curves clearly delineate contact and bulk effects, the transitional stage occuring at about 12 V. Stuart[59] has explained his measurements on Al–SiO–Al samples in terms of an electrode-limited process; the thickness of the silicon oxide ranged from 1·2 to 13·7 μm. His results are shown in Fig. 33. Below $V \simeq 70$ V the I–V characteristics are essentially thickness independent, even though there is over a decade difference between the two extreme thicknesses. Above $V \simeq 70$ V the J–V characteristics are thickness-dependent and obey a relationship of the form given by equation (35).

REFERENCES

1. Simmons, J. G., *Phys. Rev.*, **155** (1967), 657
2. Simmons, J. G., *Phys. Rev.*, **166** (1968), 912
3. Hass, G., *J. Am. Ceram. Soc.*, **33** (1958), 353
4. Ritter, E., *Opt. Acta.*, **9** (1962), 197
5. Brady, G. W., *J. Phys. Chem.*, **63** (1959), 1119
6. Faessler, V. A. and Kramer, H., *Ann. der Physik*, **7** (1959), 263
7. Simmons, J. G. and Nadkarni, G. S., *J. Vac. Sc. Tech.*, **6** (1969), 10
8. Simmons, J. G. and Nadkarni, G. S. and Lancaster, M., *J. Appl. Phys.*, **41** (1970), 538
9. Nadkarni, G. S. and Simmons, J. G., *J. Appl. Phys.*, **41** (1970), 545
10. Dresner, J. and Shallcross, F. V., *Sol. Stat. Electr.*, **5** (1962), 205
11. Mott, N., *Advan. Phys.*, **16** (1967), 49
12. Cohen, M. H., Fritzsche, H. and Ovshinsky, S. R., *Phys. Rev. Letters*, **22** (1969), 1065
13. Simmons, J. G., *J. Phys. Chem. Solids*, **32** 1987 (1971).
14. Simmons, J. G., *J. Phys. Chem. Solids* (in press)
15. Simmons, J. G., *Phys. Rev. Letts.*, **10** (1963), 10
16. Bardeen, J., *Phys. Rev.*, **71** (1947), 717
17. Price, P. J. and Radcliffe, J. M., *IBM J. Res. Dev.*, **3** (1959), 364
18. Stratton, R., *J. Phys. Chem. Sol.*, **23** (1962), 1177
19. Simmons, J. G., *J. Appl. Phys.*, **35** (1964), 2472
20. Fisher, J. and Giaever, I., *J. Appl. Phys.*, **32** (1961), 172
21. Miles, J. L. and Smith, P. H., *J. Electrochem. Soc.*, **110** (1963), 1240
22. Pollack, S. R. and Morris, C. E., *Trans. AIME*, **233** (1965), 497
23. Simmons, J. G., *J. Appl. Phys.*, **34** (1963), 1793
24. Simmons, J. G., *J. Appl. Phys.*, **34** (1963), 2581
25. Rowell, J. M., *Tunnelling Phenomena in Solids*, Eds. E. Burstein and Lundquist (Plenum Press, 1969), Chap. 27.
26. Pollack, S. R. and Morris, C. E., *J. Appl. Phys.*, **35** (1964), 1503
27. Gundlach, K. H., *Phys. Letts.*, **24A** (1967), 731

28. Gundlach, K. H. and Heldman, *Phys. Stat. Sol.*, **21** (1967), 575
29. Fowler, R. H. and Nordheim, L. W., *Proc. Roy. Soc.*, **A119** (1928), 173
30. Lenzlinger, M. and Snow, E. H., *J. Appl. Phys. 4969*, **40** 278
31. Esaki, L. and Stiles, P. J., *Phys. Rev. Letts.*, **16** (1966), 1108
32. Chang, L. L., Stiles, P. J. and Esaki, L., *J. Appl. Phys.*, **38** (1967), 4440
33. Hall, R. N., Racette, J. H. and Ehrenreich, H., *Phys. Rev. Letts.*, **4** (1960), 456
34. Wyatt, A. F. G., *Phys. Rev. Letts.*, **13** (1964), 401
35. Applebaum, J., *Phys. Rev. Letts.*, **17** (1966), 91
36. Applebaum, J., *Phys. Rev.*, **154** (1967), 633
37. Anderson, P. W., *Phys. Rev. Letts.*, **17** (1966), 95
38. Shen, L. Y. L. and Rowell, J. M., *Solid State Comm.*, **5** (1967), 189
39. Jaklevic, R. C. and Lambe, J., *Phys. Rev. Letters*, **17** (1966), 1139
40. Lambe, J. and Jaklevic, R. C., *Phys. Rev.*, **165** (1968), 821
41. Thomas, D. E. and Rowell, J. M., *Rev. Sci. Instr.*, **36** (1965), 1301
42. Schottky, W., *Physik, Z.*, **15** (1914), 872
43. Emptage, P. R. and Tantraporn, W., *Phys. Rev. Letts.*, **8** (1962), 267
44. Simmons, J. G., *Handbook of Thin Film Technology* (McGraw Hill, 1970) Chap. 14.
45. Simmons, J. G. and Nadkarni, G. S. *AC Properties of Thin Dielectric Films* (Mills and Boon, to be published).
46. Mott, N. and Gurney, R. W., *Electronic Processes in Solids* (Oxford, 1948).
47. Rose, A., *Phys. Rev.*, **97** (1955), 1538
48. Lampert, M. A., *Phys. Rev.*, **103** (1956), 1648
49. Lampert, M. A., *Reports on Progress in Physics*, **27** (1964), 329
50. Budinas, T., Mackus, P., Smilga, A. and Vivvakas, J., *Phys. Sol. Stat.*, **31** (1969), 375
51. Harth, W., *Z. Phys.*, **184** (1965), 198
52. Frenkel, J., *Tech. Phys.*, **5** (1938), 685
53. Frenkel, J., *Phys. Rev.*, **54** (1938), 647
54. Mead, C. A., *Phys. Rev.*, **128** (1962), 2088

55. Hirose, H. and Wada, Y., *Jap. J. Appl. Phys.*, **3** (1964), 179
56. Johansen, I. T., *J. Appl. Phys.*, **37** (1966), 449
57. Hartman, T. E., Blair, J. C. and Bauer, R., *J. Appl. Phys.*, **37** (1966), 2488
58. Stewart, M., *Brit. J. Appl. Phys.*, **18** (1967), 1637
59. Stewart, M., *Sol. Stat. Sol.*, **23** (1967), 595
60. Sze, S. M., *J. Appl. Phys.*, **38** (1967), 2951
61. Frank, R. and Simmons, J. G., *J. Appl. Phys.*, **38** (1967), 832
62. Schug, J. S., Willey Jr., A. C. and Lowitz, D. A., *Phys. Rev.* *B***1** (1970), 4811

INDEX

M & B MONOGRAPHS

ME/1	**Tribology** E D Hondros	0.263.51588.5
ME/2	**Electron Beam Welding** M J Fletcher	0.263.51707.1
ME/3	**Vacuum Brazing** M J Fletcher	0.263.51708.X
ME/4	**Stress Corrosion Failure** P Greenfield	0.263.51725.X
ME/5	**Glow Discharge Material Processing** R A Dugdale	0.263.51787.X
ME/6	**Diamond Grinding** J Burls	0.263.51791.8
ME/7	**Engineering Applications of Beryllium** P Greenfield	0.263.51838.8
ME/8	**Whiskers** C C Evans	0.263.51844.2
CE/1	**Microbiological Corrosion** G H Booth	0.263.51584.2
CE/2	**Solvent Treatment of Coal** W S Wise	0.263.51706.3
CE/3	**The Fluidised Combustion of Coal** D G Skinner	0.263.51714.4
CE/4	**Carbonisation of Coal** J Gibson D H Gregory	0.263.51774.8
CE/5	**The Principles of Gas Extraction** P F M Paul W S Wise	0.263.51802.7
CE/6	**Metal-Air Batteries** D P Gregory	0.263.51712.8
CE/7	**Fuel Cells** D P Gregory	0.263.51713.6

EE/1	**Threshold Logic** S L Hurst	0.263.51587.7
EE/2	**The Magnetron Oscillator** E. Kettlewell	0.263.51641.5
EE/3	**Material for the Gunn Effect** J W Orton	0.263.51773.X
EE/4	**Microcircuit Learning Computers** I Aleksander	0.263.51724.1
EE/5	**DC Conduction in Thin Films** J G Simmons	0.263.51786.1
EE/6	**Reading Machines** J A Weaver	0.263.51839.6

Forthcoming titles

Creep of Metals	P Greenfield
Cavitation	I S Pearsall
Magnesium	P Greenfield
Plasma Spraying	R F Smart and J A Catherall
Biotechnology of Industrial Water Conservation	B E Purkiss
Speech Synthesis	J N Holmes
Glow Discharge Display	G F Weston
Superconductivity	I M Firth
Ferrites	E E Riches

M & B TECHNICAL LIBRARY

TL/ME/1	**Hydrostatic Extrusion** J M Alexander B Lengyel	0.263.51709.8
TL/ME/2	**The Heat Pipe** D Chisholm	0.263.51711.X
TL/CE/1	**Boron** A G Massey J Kane	0.263.51775.6
TL/EE/1	**Linear Electric Motors** E R Laithwaite	0.263.51586.9
TL/EE/2	**An Introduction to the Josephson Effects** B W Petley	0.263.51705.5

Mills & Boon publish monographs covering a wide range of topics in the three engineering disciplines. Further details may be obtained on request.

Mills & Boon Limited

17–19 Foley Street

London W1A 1DR